City Water Matters

Sophie Watson
City Water Matters
Cultures, Practices and Entanglements of Urban Water

palgrave
macmillan

Sophie Watson
Sociology Department
Open University
Milton Keynes, UK

ISBN 978-981-13-7891-1   ISBN 978-981-13-7892-8 (eBook)
https://doi.org/10.1007/978-981-13-7892-8

© The Editor(s) (if applicable) and The Author(s), under exclusive licence to Springer Nature Singapore Pte Ltd. 2019
This work is subject to copyright. All rights are solely and exclusively licensed by the Publisher, whether the whole or part of the material is concerned, specifically the rights of translation, reprinting, reuse of illustrations, recitation, broadcasting, reproduction on microfilms or in any other physical way, and transmission or information storage and retrieval, electronic adaptation, computer software, or by similar or dissimilar methodology now known or hereafter developed.
The use of general descriptive names, registered names, trademarks, service marks, etc. in this publication does not imply, even in the absence of a specific statement, that such names are exempt from the relevant protective laws and regulations and therefore free for general use.
The publisher, the authors and the editors are safe to assume that the advice and information in this book are believed to be true and accurate at the date of publication. Neither the publisher nor the authors or the editors give a warranty, express or implied, with respect to the material contained herein or for any errors or omissions that may have been made. The publisher remains neutral with regard to jurisdictional claims in published maps and institutional affiliations.

Cover illustration: Nigel Howard / Alamy Stock Photo

This Palgrave Macmillan imprint is published by the registered company Springer Nature Singapore Pte Ltd.
The registered company address is: 152 Beach Road, #21-01/04 Gateway East, Singapore 189721, Singapore

For Jessie

Acknowledgements

This is a book that began some years ago, taking shape in fits and starts over the following years. It was very much a sole endeavour but there are several people to thank. First, to the people who gave their time to be interviewed and who seemed to care about water, in its many guises, as much as I did. Second, the book came together during a month's visiting fellowship at the Max Planck Institute in Gottingen; my thanks to Steve Vertovec for inviting me there. My thanks also to those who read the first chapter and made me realize I had a book worth publishing, to Paddy Hillyard for first encouraging me to think about water, to Jeri Johnson for introducing me to *Ulysses*, to Edward Wigley for his interviews on Wudu, to Russell Hay for his idiosyncratic contributions and cuttings, to Olly Zanetti for his enthusiasm, and to Farhan Samanani for his very insightful comments on the final draft. The earth without water, and all who live on it, will not survive. As climate change activists put it: *The time to act is now.*

Contents

1 City Water Matters: Cultures, Practices and Entanglements of Urban Water—An Introduction   1

2 Public Water Features: Assembling Publics, Enlivening Spaces, Promoting Regeneration   15

3 Consuming Water: Habits, Rituals and State Interventions   43

4 River Powers: Assembling Publics, Connections and Materials in a Global City   73

5 Embodied Water Entanglements: Sex/Gender, Race/Ethnicity and Class Urban Sanitation Practices   105

6 Public Waters: The Passions, Pleasures and Politics of Bathing in the City   135

7 Differentiating Water: Cultural Practices and Contestations   167

| 8 | Water Traces in Urban Space | 197 |
| 9 | Final Word | 215 |

List of Figures

Photo 2.1	Granary Square Kings Cross. Sophie Watson	35
Photo 2.2	A drinking fountain in Venice. Sophie Watson	39
Photo 3.1	Green doctor anonymous. Sophie Watson	57
Photo 4.1	A group of Watermen	78
Photo 4.2	A waterman's licence. Sophie Watson	79
Photo 4.3	The waterman's coat. Sophie Watson	94
Photo 5.1	A council laundry—early 1900s. Sophie Watson	123
Photo 5.2	A wash and fold laundry in New York. Sophie Watson	125
Photo 5.3	Sycamore laundry boxes. Sophie Watson	130
Photo 6.1	In de Arte Natandi. Sophie Watson	139
Photo 6.2	The report of the British Spas Federation. Sophie Watson	144
Photo 6.3	Hampstead Women's Pond. Katrina Silver	153
Photo 6.4	The Bondi Icebergs Pool. Sophie Watson	162
Photo 7.1	Mosque washroom. Sophie Watson	172
Photo 7.2	Wudhu washing instructions. Sophie Watson	173
Photo 7.3	The river Ganges at Varanasi. Jessie Watson	188
Photo 8.1	A horse trough in Sydney. Sophie Watson	200
Photo 8.2	An African water tap	201
Photo 8.3	The Wapping Old Stairs	202
Photo 8.4	A cistern in Rye, Sussex. Sophie Watson	205
Photo 8.5	The Acquedotto in Puglia. Jessie Watson	207
Photo 8.6	Mission Brewery San Diego	210
Photo 8.7	The Bondi sewage outflow plant. Sophie Watson	211

1

City Water Matters: Cultures, Practices and Entanglements of Urban Water— An Introduction

In *Ulysses* (Joyce, 1922) what was it about water that Leopold Bloom admired? It was its

> universality: its democratic equality and constancy to its nature in seeking its own level… its violence in seaquakes, waterspouts, artesian wells, eruptions, torrents, eddies, freshets, spates, groundswells, watersheds, waterpartings, geysers, cataracts, whirlpools, maelstroms, inundations, deluges, cloudbursts: its vast circumterrestrial ahorizontal curve: its secrecy in springs, and latent humidity,… the simplicity of its composition, two constituent parts of hydrogen with one constituent part of oxygen: its healing virtues… its infallibility as paradigm and paragon: its metamorphoses as vapour, mist, cloud, rain, sleet, snow, hail: What in water did Bloom, waterlover, drawer of water, watercarrier returning to the range, admire? Its strength in rigid hydrants: its variety of forms in loughs and bays and gulfs and bights and guts and lagoons and atolls and archipelagos and sounds and fjords and minches and tidal estuaries and arms of sea: its solidity in glaciers, icebergs, icefloes: its docility in working hydraulic millwheels, turbines, dynamos, electric power stations, bleachworks, tanneries, scutchmills: its utility in canals, rivers, if navigable, floating and graving docks: its potentiality derivable from harnessed tides or watercourses falling from level to level: its submarine fauna and flora (anacoustic, photophobe) numerically, if not literally, the inhabitants of the globe. (James Joyce's *Ulysses*. Episode 17. Ithaca)

Human beings are constituted of water. All our organs are made up of different amounts of water—the brain, lungs and kidneys contain 85% water, the bones—31%. We not only depend on water, without water we would die. Water lies at the very heart of the interconnectedness and entanglements of humans with our environment and reveals, arguably more than any other substance, the impossibility of thinking of ourselves as separate from nature. Water is far from being a naturally occurring terrestrial resource but is shaped in multiple ways by human activities and cultural practices. Water exists as a resource through a complex intersection of socio-technical networks and systems and is a site of different cultural meanings and social practices across time and space. Water is both a source and a force, as the 2018 tsunami in Indonesia so graphically illuminated. Water elicits a set of technologies for containing it and directing it; it is enmeshed in a myriad of governmental and regulatory practices as well as private markets and complex forms of provision. Its abundance as well as its scarcity, though products of technology, is equally constituted in public discourses and political decisions, implicated in relations of power as these are so-called a 'natural' occurring phenomena such as rains, floods and drought. Water enables and assembles a multiplicity of publics, embodied practices, cultural practices and diverse forms of sociality. Water is far from being just the natural resource it is often assumed to be.

Water stretches and flows across human–non-human networks (Allon 2009). Nowhere are the effects of human activities on the Earth's ecosystems, denoted by many as the Anthropocene, more visible. Not only does water cross the boundaries of different substances, with complex intersections and effects, the very complexities of current and impending water crises across the world, lead to greater uncertainties as to what kinds of solutions are possible. Indeed, many see water as the most pressing concern of the twenty-first century, which is likely to lead to massive migrations and water wars. At the same time, there is less and less certainty as to how best to resolve the scarcity of water, or its overabundance in the form of floods or rising sea levels in many parts of the world.

Water is never far away from the centre of life and thought. Different civilizations at different times have intervened in the flow of water and its provision, from the Ancient Mayan to the vast hydrological systems of the Roman Empire. Artists, engineers and scientists have been intrigued by its workings and powers. Leonardo da Vinci in his famous Codex Leicester (1506–1510)—bought by the Bill Gates foundation for USD 30.8 million twenty years ago—muses on the role of water in shaping the world, sketching its ebbs and flows and constructing practical schemes for its management.

For da Vinci 'we may say that the earth has a spirit of growth and that its flesh is the soil… its blood the veins of its waters. The lake of the blood that lies around the heart is the ocean. Its breathing is by the increase and decrease of the blood and its pulses, and even so in the earth is the ebb and flow of the sea'. It would be a mistake to think that contemporary social theorists were the first to posit the human/non-human space where water resides.

Water is theorized in many different registers and through many different frames: crisis, infrastructure, symbol, culture, politics, management and delivery, consumption, the economic and the social. Each of these spheres are interconnected and related and not easily disentangled. Water is emblematic of the powerful interconnections between human/non-human, and nature and culture, where these entanglements are in a constant process of transforming cityscapes and landscapes, which in turn produce new waterscapes and manifestations of the 'natural world'. Water has the capacity to make things happen, to bring new socialities and publics into being. Water is an intrinsic part of everyday life, often invisible in its workings and taken for granted, only entering public discourse and visibility when it becomes a matter of concern through, for example, its scarcity, or its potential for danger or for the accumulation of profit. Water is deeply political, implicated in relations of power and constitutive of social, cultural and spatial differences. Water is highly contested both as a resource and as a site of complex meanings.

Though the impending crisis of water scarcity is likely to be the greatest challenge of the twenty-first century, this is not the central thrust of this book—there are others far more expert and knowledgeable to make this case. The aim of this book is more modest, though it too has its political thrust. Here I follow in the footsteps of those who have argued for the importance of water as a cultural object, and as a source of complex meanings in everyday life, whether it be Mumbai or London, Hanoi or Paris, Sydney or Los Angeles. Humans need water to thrive, and many daily practices and habits that often go unnoticed are connected to the presence of water. Water is needed for bodies to keep clean, for embodied pleasures, for ritual practices, for the beautification of urban sites and so on, and each of these are enabled by the socio-technical systems and structures of a specific historical and material context. This book then brings cultural practices to the fore, arguing also for their embeddedness in a wider social, political, economic and technological context. In the next section, I briefly sketch some of the extensive research and writing on water, before outlining the strands and arguments that underpin this book.

Crisis

The current crisis of water is framed in two ways—there is too little and there is too much.

Climate change brought about by human activity is the major culprit in both cases, with wetter regions typically becoming wetter and drier regions becoming drier. It is estimated that half the world's population (3.6 billion) live in areas that are potentially water scarce for at least one month of the year (WWAP 2018: 3). The global demand for water has been increasing by 1% annually due to population growth. Water pollution has worsened in the majority of rivers across Africa, Latin America and Asia, and there has been increasing ecosystem degradation due to the poor condition of soil resources with impacts on higher evaporation and erosion, and the degraded state of the world's forests. Importantly water scarcity is unevenly distributed.

There is a tendency in much of the literature towards pessimistic discourses. But as the World Water Council point out 'the crisis is not about having too little water to satisfy our needs. It is a crisis of managing water so badly that billions of people—and the environment suffer badly' (WWC 2018). The WWC makes a crucial point, which is that as long as we do not ourselves face water scarcity, we believe that the taps will keep on flowing. Huge quantities are wasted through water-intensive industrial and agricultural practices (which account for 20% and 70% of global water withdrawals respectively) and through individual practices. For the former two, greater regulation is required; while for the latter this is to a large extent a question of habits and everyday consumption practices, which are hard to change. Peoples' habits and practices are an area of current intervention by water authorities, as we see in Chap. 3. But there are also cultural shifts at a wider level that matter. One significant arena is that of food, where an increasing understanding of the quantities of water required to produce beef—1 kg requires 1300 litres, in contrast to potatoes—1 kg requires only 100 litres of water—has led considerable numbers of people, particularly younger people, to adopt vegetarian or vegan diets. Ironically, another trend towards substituting cow's milk with almond milk has deleterious effects where water is concerned. 1 litre of almond milk requires more than 2 litres of water, yet 80% of the world's almond crop comes from California which is beset by droughts. Scarcity leads to tensions between different users and trans-boundary conflicts and to migrations of populations to water-rich areas. Some commentators predict water wars in the coming decades. Water scarcity is highly political and distributed unequally across nations constituted in relations of power. Deforestation affects some parts of the world and large multinationals run roughshod over resistances from powerless groups, often indigenous populations who for centuries owned and managed their lands for centuries with environmental awareness.

Equally a consequence of climate change are rising sea levels and flooding, with severe impacts on coastal cities from Miami to Mumbai. Jeff Goodall (2017) provides a stark analysis in his book *The Water will Come; Rising Seas, Sinking Cities and the Remaking of the Civilised World*. The collapse of the ice sheets in the Antarctic and Greenland at a greater rate than previously thought are leading to predictions of rising sea levels of 8 feet by 2100—a figure raised to 11 feet if carbon emissions continue at present levels. According to Goodall, the difference of a three feet sea rise and a rise of six feet is the difference between a wet but liveable city and a submerged city. Major infrastructures like New York's JFK airport or coastal nuclear reactors are likely to be underwater in 100 years. Yet it is the 145 million people who live three feet above sea level who are the most likely to be affected. This too is a highly politicized crisis, which will affect many people in the Global South, creating generations of climate refugees and making the current refugee crisis seem like a drop in the ocean (to deploy a water metaphor). Similarly, the disastrous effects of flooding in cities of the Global North are more likely to affect poorer urban populations with limited resources, as the rupture of the levees in New Orleans so clearly demonstrated. As Goodall points out, wealthy people will move or elevate their homes or build sea walls to surround their gated communities.

These narratives of disaster can be disempowering though resulting in a response of despair and passivity. Even Goodall suggests a sharing of resources and know-how to build more adaptive city environments, while national and international water organizations provide clear policy initiatives which could at the very least halt this trend towards disaster. Local communities are increasingly mobilized to devise new ways to manage water at the local level. The UK Water Partnership (2015) report on the 'Future of Cities', for example, proposes new urban design initiatives to raise city areas: improved city groundwater management, shifting water-related habits, decentralized urban infrastructures and groundwater management while larger-scale directives and reports by international organizations like *United Nations Water* consistently draw attention to water issues in their reports.

Water Resources and Infrastructures

Geographers have eloquently drawn attention to water flows and infrastructures as physical and material landscapes as well as landscapes of power. Swyngedouw (2004) argues that the water supply relies on constant mastering and transforming of 'natural water' to the extent where it is impossible to imagine the provision of water outside of the large bureaucratic and engineering control systems which enable centralized decision making and monopoly control where profits can be extracted (Swyngedouw 2004: 1). This Promethean modernist project disrupts our imagination of water delivered from the tap at home, as being connected to the vast technological shrines of the dams and reservoirs, often distances away (Kaika 2005: 3). Such a connection only becomes evident when the water supply is threatened. As Kaika (ibid.: 4) argues, the project of modernity was based on 'establishing intricate flows of natural elements, social power relations and capital investment cycles—which not only didn't separate nature from the city, but instead wove them together more closely into a socio-spatial continuum'. For both these writers, and also Gandy (2004) the flow of water, the flow of power and money are inextricably linked, which have the potential to encourage greater social cohesion or generate new forms of political conflict and social cleavages.

This connects with struggles over the control and ownership over water authorities, which rage across the globe. Where once water was seen as a public good, as a basic human right, with privatization, it has shifted into being seen as a source for the extraction of profit, moving from the realm of provision by national governments to the control of global financial markets and national and international banks. In this context, discourses of nature as unpredictable are mobilized as the source of the crisis and scarcity of water, thus constructing it as a valuable and precious commodity, best managed by the market economy where increases in water prices become a legitimate and inevitable response (Allon 2009).

Thames Water, the largest UK water authority, is illustrative of the debacles that have often ensued. In 2006, the Australian Macquarie bank bought the company. Between 2007 and 2010, it borrowed GBP 2 billion which was used for the benefit of the bank and its investors (Robinson 2017), contravening the conditions laid down by the regulator. Under Macquarie's control, the total returns made by the bank and its investors from Thames Water averaged between 15.5% and 19% a year, twice the expected amount. At the same time, Thames was consistently criticized for its widespread leakages across the region, for its extensive pollution of the Thames and other rivers with untreated sewage between 2012 and 2014. A judge at the Aylesbury Crown Court said there had been 'inadequate investment, diabolical maintenance and poor management'. In the following five years, environmental performance as measured by the Environment Agency was said to have improved significantly, though complaints continued to be rife. In 2017, Macquarie sold its final stake in Thames Water for an estimated GBP 1.35 billion to the infrastructure arm of the Canadian pension fund Omers and the Kuwait Investment Authority, highlighting a growing interest in shares in British utilities which are increasingly seen as 'prized assets' since they deliver steady returns (Plimmer 2017). This is a story about water as money, not water as part of the commons.

Post-humanist Approaches

In recent years a post-humanist approach to thinking about water has emerged, which seeks to redefine the place of humans in the world where the human is seen as one form of life amongst many. This approach challenges binary thinking—nature/culture, human/animal proposing the notion of *natureculture* as a synthesis of nature and culture which posits their inseparability in ecological relations which are both socially, biologically and physically formed. Thus, biology and socio-culture are mutually constitutive, radically producing and shaping one another. In this frame, the world is conceptualized as open, immanent and ever-becoming and humans cannot be separated from water but are constituted by, and constitutive of its very substance, its flow and fluidity. In this vein, scholars write of 'wet ontology' marine biology, surfing and so on as processes where humans and water are thoroughly interconnected. Thus, Steinberg and Peters (2015: 247) argue that the ocean is an ideal spatial foundation for moving away from notions of fixity 'since it is indisputably voluminous, stubbornly material, and unmistakably undergoing continual reformation, and that a 'wet ontology' can reinvigorate, redirect, and reshape debates that are all too often restricted by terrestrial limits'. In a related move, drawing on French theorists such as Irigaray, Deleuze and Merleau-Ponty, Astrida Neimanis (2017) takes the interconnectedness of water and humans into post-human feminist phenomenology seeing our bodies as being fundamentally part of the natural world and not separate from or privileged to it.

Culture, Everyday Life and Symbolic Meanings

A fourth strand of writing engages more with water as a cultural practice or object. Allon and Sofoulis (2006) make the distinction between everyday water and big water. Everyday water focuses on the cultural meanings and water practices embedded in daily life—such as washing, cooking, bathing, gardening and drinking which function in interaction with the technologies and socio-technical systems that constitute domestic dwellings. They explored everyday water through a research project which looked at the values, practices and interactions around water amongst households in Western Sydney. As we see also in Chap. 3, they concluded that complex meanings and values influence patterns of domestic water consumption, such as time spent in the shower or watering the garden. Big water, on the other hand, concerns large-scale projects which involve the mastery of nature in the construction of giant reservoirs and catchment areas, dams, pipe networks, pumps and so on, all of which attest to the grand engineering projects of modernity. Much of this writing is embedded also in socio-technical accounts and imaginaries which see infrastructures as actively engaged in producing specific forms of water policies and developments. Molden et al. (2016), for example, explore the role of traditional stone waterspout technologies in Nepal as a visual discourse of materials, practices and texts, finding that the stone spouts represent material and symbolic sites for residents to express discontent with hegemonic visions of modernization and make claims on urban space.

Drawing on ethnographic studies of particular places, Strang's extensive writing (2006, 2009) on water takes a similar approach in drawing attention to the diverse meanings of water. In her ethnographic research in Dorset, Strang (2006) explores how human engagements with water, in particular with rivers and water supply infrastructure—mediate individual, familial and wider collective identities in what she sees as a shifting cultural 'fluidscape' of social, spatial, economic and political relationships (2006). In each of her different studies Strang highlights water as a material substance, which quite literally constitutes what it means to be human and which acts as a 'natural symbol of sociality and interconnection, and human-environmental interdependence and interconnectedness'. In her book *Gardening the World* (2009) she focuses on the ways in which the different beliefs and practices, for example of industrial versus domestic users, lead to different forms of environmental engagement, some of which are more socially and economically sustainable than others. Despite such variations, she suggests that water has 'core meanings that recur cross culturally, standing symbolically for life, wealth and health, and most particularly for spiritual, social and ecological regeneration' (30). Not surprisingly, like others (Swyngedouw 2004) this leads her to a critique of the privatization of water which she sees as a common good and as an integral part of the ecological order.

Strang is not alone in thinking about water as a symbol and source of complex and contested meanings. A research group shaped by Syse and Oestigaard (2010) spent some years exploring the role of water in history and development, suggesting three interconnected layers of the physical form, human modifications and adaptations and the cultural concepts and ideas of water and water systems. Following a workshop on 'Fluid Approaches to History' in 2009, these authors collected together an anthology (Syse and Oestigaard 2010) discussing the relationship between the scientific and technological developments and images and ideas about water between 1500 and 1850, thus connecting the scientific and cultural arenas in a way that is often absent in water studies. 'Thinking with Water' (Chen et al. 2013) similarly seeks to analyse the culture of water and to consider the different ways we interact with and imagine water, underlining water as a political, auditory, material, visual, postcolonial and ethical subject.

Cleary then there is a shift from the dominance of a way of thinking which focuses on water in terms of hydro-engineering, hydraulics or vast infrastructures—the mastery of water in the systems of 'big water' (Allon 2009), to approaches which see water as networked, interconnected, symbolic and cultural, entangled in a complexity of socio-technical and material assemblages which constitute everyday life. Nevertheless, water as a private commodity exploited for profit, rather than a common good, underpins its delivery and management in most countries. As such, it will continue to produce social/spatial divisions and differences across the globe and within communities, based on gender, class, ethnicity/race which are only likely to worsen as climate change propels us towards an ever more dangerous precipice. On 8 October 2018 the UN's Intergovernmental Panel on Climate Change released a report calling for urgent action to phase out fossil fuels reaffirming the challenge to limit global warming rises to 1.5 °C—without which greater flooding and other environmental disasters are inevitable.

This book draws its influences from all these strands, yet most centrally argues for the importance of water as a cultural object, and as a source of complex meanings and practices in everyday life, embedded in the socioeconomics of local water provision. What I am interested in therefore are human-organized systems, meanings and practices, in a way that does not neglect the materiality of water and water–ecological relations, but which does not take these as its primary focus. Water, precisely because of its fluidity, in some sense resists definition, in that, however much human infrastructures and representations strive to contain and channel water, water continually 'leaks' out of this containment. Because it is a highly fluid substance that is difficult to contain, physically, and because it is so essential to all life, particular cultural representations can never quite monopolize it. Water is thus difficult to encapsulate, connected in different ways and spaces to unwanted fluidities—of migrants in the case of crisis, of gender or sex, of bodies and identities.

The book itself is thus also fluid—breaking out of traditional categories and exploring, in a serendipitous way, different aspects of water as it settles or is unsettled in cities. Each city is different, though there are some commonalities between global cities or between mega cities of the Global South, for example. This book mainly explores London, although other cities make their mark. Water makes and unmakes cities across the ages, enabling and disabling daily life as it ebbs and flows into domestic and public spaces. Yet water is often ignored, invisible or forgotten as vibrant matter in the city. The book can then be read in no particular order, it can be dipped into or out of, just like water in a pond or the sea. Each chapter aims to capture one element of water's fluid existence in the world, as material object, cultural representation, as movement, as actor, as practice and as ritual. In Chap. 5 ('Embodied Water Entanglements: Sex/Gender, Race/Ethnicity and Class Urban Sanitation Practices') I explore the embodied and material practices associated with laundry, washing and toilets historically and to the present day.

Another strand is water's capacity to assemble publics. Through and around the medium of water different people and institutions come together, be it for work, pleasure, movement, cultural practices, consumption or governance. During the nineteenth century some 40,000 men earned a living on or about the Thames (Stow 1876): lightermen to carry passengers, dockers to unload the ships, watermen to carry the goods ashore, mudlarkers sifting the banks to find treasures, sewer hunters delving in the drainage for old coins, bargemen to carry large loads along the river, whisky men to quench the worker's thirst. These were just some of the men drawn to the power of the river to enable work and to bring people to its waters. This is a story that repeats itself in urban rivers everywhere, illustrated here by the Thames (Chap. 4 'River Powers: Assembling Publics, Connections and Materials in a Global City'). Not only does water assemble working people, through its capacity to provide embodied pleasures, as a site of immersion, relaxation and exercise, water everywhere engages people in aquatic pursuits, enabling healthy and happy bodies, soothing troubled souls and connecting individuals across their differences as they share spaces in often unexpected and unplanned encounters. 'Public Waters: The Passions, Pleasures and Politics of Bathing in the City' provides a window onto the multiple publics that use lidos, ponds and baths in cities. In a similar vein, public water features in the city bring together urban designers, water engineers and multiple publics to gaze and more recently to actively engage, as I explore in Chap. 2 'Public Water Features: Gathering Publics, Enlivening Spaces, Promoting Regeneration'. Publics then are heterogeneous. This is very clear when considering how water is imbued with religious meanings and symbols and is core to a plethora of religious practices and rituals, which often go unnoticed until they are contested. 'Differentiating Water: Cultural Practices and Contestations' thus constitutes a key chapter of the book. Finally understanding the practices and habits associated with consumption, which are differentiated by age, gender and ethnicity, is crucial in strategies to reduce household use of water; this forms the basis of the discussion in Chap. 3 'Consuming Water: Habits, Rituals and State Interventions'.

The specificity of water sites and water practices, then, are core to the concerns of this book. Underpinning each of these are several key themes. First are the interconnectedness of humans and non-humans, of nature and culture, and the complex entanglements of water in all its many forms. The second argument is that water constitutes multiple differences which are themselves not fixed, but which shift and change across time and place. Third, that water is implicated in relations of power, often invisible, but present nevertheless in the workings of daily life in all its rhythms and forms. Thus also, differences are themselves connected to power, and water sites and resources mark boundaries and borders, and are political and contested. And finally, water has the capacity to assemble a multiplicity of publics and constitute new socialities and connections. The particular sites and water cultures explored here (due to limited resources) are in London, though other cities find a place. Further work needs to be done to develop some of the rich work on the very specificity of water cultures in different cities across different socio-political and socio-technical environments, to highlight how context plays such a major part in how water is lived in practice across space and time. As the history of Angkor Wat attests so vividly, cities, and their inhabitants, without water will die, and so will their cultures.

References

Allon, Fiona. 2009. Water. In *International Encyclopedia of Human Geography*, ed. R. Kitchin and N. Thrift, vol. 12, 210–206. Oxford: Elsevier.

Allon, Fiona, and Zoe Sofoulis. 2006. Everyday Water: Cultures in Transition. *Australian Geographer* 37 (1): 45–55.

Chen, Cecilia, Janine Macleod, and Astrida Neimanis. 2013. *Thinking with Water*. McGill-Queen's University Press.

Gandy, Matthew. 2004. Rethinking Urban Metabolism: Water, Space and the Modern. *City* 8 (3): 363–379.

Goodall, Jeff. 2017. *The Water will Come; Rising Seas, Sinking Cities and the Remaking of the Civilised World*. New York: Little, Brown Hachette Book Group.

Kaika, Maria. 2005. *City of Flows Modernity, Nature and the City*. London: Routledge.

Molden, Olivia, Nicholas Griffin, and Katie Meehan. 2016. The Cultural Dimensions of Household Water Security: The Case of Kathmandu's Stone Spout Systems. *Water International* 41 (7): 982–997.

Neimanis, Astrid. 2017. *Bodies of Water: Posthumanist Feminist Phenomenology*. London: Bloomsbury.

Plimmer, Jill. 2017. Macquarie Sells Final Stake in Thames Water for @1.35 bn. *Financial Times*, March 14.

Robinson, Michael. 2017. How Macquarie Bank Left Thames Water with Extra £2bn debt. *BBC News*, September 5.

Steinberg, Philip, and Kimberley Peters. 2015. Wet Ontologies, Fluid Spaces: Giving Depth to Volume through Oceanic Thinking. *Environment and Planning D: Society and Space* 33 (2): 247–264.

Stow, John. 1876. *A Survey of London: Written in the Year 1598*. London: Chatto and Windus.

Strang, Veronica. 2006. Substantial Connections: Water and Identity in an English Cultural Landscape. *Worldviews: Global Religions, Culture, and Ecology* 10 (2): 155–177.

———. 2009. *Gardening the World: Agency, Identity and the Ownership of Water*. New York: Berghahn Books.

Swyngedouw, Erik. 2004. *Social Power and the Urbanisation of Water*. Oxford: Oxford University Press.

Syse, Karen Lykke, and Terje Oestigaard, eds. 2010. *Perceptions of Water in Britain from Early Modern Times to the Present: An Introduction*. Bergen: University of Bergen.

UK Water Partnership. 2015. *Future of Cities*. Natural Environment Research Council.

World Water Council. 2018. Water Crisis. http://www.worldwatercouncil.org/en/water-crisis.

WWAP (United Nations World Water Assessment Programme)/UN-Water. 2018. *The United Nations World Water Development Report 2018: Nature-based Solutions for Water*. Paris: UNESCO.

2

Public Water Features: Assembling Publics, Enlivening Spaces, Promoting Regeneration

Fountains, their form and function, offer a sparkling window into the cultural, symbolic and spiritual life of different cities at the time of their construction. Fountains have been created to meet a diversity of purposes: to enact power, to beautify, to memorialize subjects, to quench thirst, to enable bathing and for religious practices—often to wash away sins or symbolize deities (Barber 2003). More recently, they have been created to enliven public space, to regenerate cities and communities and to promote cities on the international stage. Writing about Roman fountains Morton (1970: 17) rather touchingly has this to say: 'the fountain has no enemies, even in cold countries: it is a device or invention which has given nothing but pleasure in the course of a long history'. For many centuries fountains were public objects of decoration in the city, largely separated from publics by hard boundaries, which discouraged engagement between the water and its onlookers, who instead were enrolled in practices of admiration and wonder, rather than mutual interaction and play. Indeed people were often forbidden to enter the water and the borders were (and continue to be) policed. Increasingly, however, water is integrated into the urban fabric where fountains assemble water and bodies in spaces of mutual encounter, engagement and play.

Fountains not only offer a lens to understand urban life and its textures, they also reflect the prevailing understandings of hydrological engineering and design, which in some parts of the world, as we shall see, revealed a highly sophisticated knowledge of the physical and material world many centuries ago. Most often fountains were enabled by gravity, drawing on a water source such as a dam or aqueduct at a higher level, but in recent years, many fountains are operated digitally. Istanbul, Paris and Rome each offer a rich story of the different uses, architectures and representations, of the fountains in the city, and the technologies that underpinned them. Having taken a brief journey to these cities to reveal the marvel of water technologies deployed and the intentions implicit in their form, I then explore fountains in the contemporary city, and their role in animating public space and regeneration, memorialization, place making, international city branding and the quenching of thirst.

Rome: Enacting Power

From the inception of cities, fountains have celebrated and publicly established, local or national powerful figures on the one hand, or consolidated city power on the other. Of all the European cities, Rome is best known for its fountains, where the power of the church and state is materialized in its public water features. The Four Rivers fountain in Piazza Navona represents one such illustration. Designed by Bernini for Pope Innocent X Pamphilj, who owned a palace on the square and inaugurated in 1651, the fountain represents the four rivers that stand for the four continents known at that time. According to tradition, Bernini carved the arm of the statue raised to protect itself from the imminent collapse of the church that had been enlarged and reconstructed by his great rival Borromini.

A deep and indelible relationship binds Rome with its waters, where the abundance of water led the writers of antiquity to assign the city the name Regina Aquarum (Venturi and Sanfilippo 1996: 17). The Tiber—the Pater Tiberinus, was the very symbol of the city and was celebrated each year in December, and often the river is represented together with the twins Romulus and Remus, underlining the close ties between the city, its mythical founders and the river (ibid.: 18). The hydrological and engineering knowledge that underpinned the Roman water system remains revered to this day and its traces endure in the stones and architectures of the contemporary city. The first aqueduct—the Acqua Appia was constructed in 312 BC to bring water from the Lucullen estate to the city. By 226 AD when the last aqueduct—Acqua Alexandrina was built, there were 11 aqueducts, between 11 and 56 miles long, pouring the water of springs, a river and a lake into the city of Rome. More water was brought to the city than was needed for the domestic use of the population since many aqueducts were built to satisfy the Roman love of the bath. By 410 AD, there were 1212 public fountains, 11 great imperial thermae and 926 public baths (Morton 1970: 28–31), which embodied the sociality and public lifestyle of contemporary Roman urban culture. Never before had a city known such a display of water leading Pliny (quoted in Forbes 1899: 3) to marvel:

> If we only take into consideration the abundant supply of water to the public, for baths, ponds, canals, household purposes, gardens, places in the suburbs, and villas, and then reflect upon the distances that are traversed, the arches that have been constructed, the mountains that have been pierced, the valleys that have been filled up, we must of necessity admit that there is nothing to be found more worthy of our admiration throughout the whole universe.
>
> And as Galen the Greek doctor who visited the city in 164 AD, remarked, not only were the fountains beautiful but also 'none emits water that is foul, mineralized, turbid, hard or cold.

By the fourth century following the collapse of the Roman Empire, the barbarian ravages, major floods and the disappearance of technological skills, the water system had ceased to function. Over ten centuries passed before the problem of inefficient water supply re-entered the political arena as a cause for concern. In the fourteenth century, Pope Nicholas V (1397–1455), decided to embellish the city and to rebuild the ruined Acqua Vergine that had brought clean drinking water to the city from eight miles away. Two centuries later, the pontificate of Gregory XIII (1572–1585)—remembered by his contemporaries as the fountain pope—saw the renewed construction of fountains to provide townsfolk with a principal source of water (Morton 1970: 42). By the seventeenth and eighteenth century following the reconstruction of the aqueducts under a succession of Popes, a wealth of fountains were built across the piazzas and streets of the city, adorned with elaborate sculptures and mythological figures, which remain one of the defining features of the contemporary city—it is estimated that there are some 2000 today. Once only operating by gravity, the fountains now also draw their water from mechanical pumps; some are embellished while others are simple free-standing cylindrical drinking fountains, given the nickname 'nasoni' (big noses) due to the shape of their protruding taps.

These 2800 'nasoni' are as emblematic of the city as the Coliseum, having been initially installed in 1874 to provide clean, easily accessible drinking water for Romans and for street markets selling fruit and vegetables, flowers, fish and meat. Each spout is punctured by a tiny hole enabling drinkers to block the end of the spout with their hand, causing a thin stream of water to arch out of the hole, enabling drinkers to quench their thirst without bending low. The water is fresh and cold and goes under the nickname 'l'acqua del sindaco'—the mayor's water. Most of the cast-iron fountains are stamped with the letters SPQR—Senatus Populus Que Romanus, or the Senate and the People of Rome, the symbol of the ancient city (Squires 2017). In the summer of 2017, the fountains were turned off due to a severe drought across Italy. The head of the utility company Acea that manages the fountains was well aware of the 'inconvenience that this will cause, but it is due to the exceptional drought,' but the response was necessary he argued, to relieve the pressure on volcanic Lake Bracciano, north of the city, one of its principal water sources, where the level had been falling for some weeks. The decision came under fierce critique from locals and consumer groups, who suggested that the move would force tourists and citizens to buy bottles of water in bars and shops pushing up prices. Only 50% of wasted water was represented by fountain use, they argued, which like in London 50% was caused by pipeline leaks (Squires 2017). Turning off Rome's fountains, it was argued, would give Acea a chance to carry out much-needed improvement works on the piping, according to Saccani, who said old pipes would be replaced and leaks repaired throughout the city (Edwards 2017).

In a gesture of solidarity with the city, the Vatican followed suit by turning off 100 of its fountains. According to the Vatican spokesman, Greg Burke, it was the first time that authorities in the spiritual home of the world's 1.2 billion Catholics could remember being forced to turn off the fountains (Guardian 2017). Standing in St Peter's Square, where two fountains by seventeenth-century sculptors Carlo Maderno and Gian Lorenzo Bernini stood dry, Burke said the Vatican wanted to help: 'This is the Vatican's way of living in solidarity with Rome, trying to help Rome get through this crisis,' he told Reuters TV (Guardian 2017). The decision reflected the ecologically minded concerns of the current Pope Francis, who demanded swift action by world leaders in the first ever papal document dedicated to the environment (Guardian 2017).

In recent years, the image of the purity of public water has been sullied. A study in Italy by the consumer organization Altroconsumo (The Local 2015) analysed the drinking water from public fountains across 35 cities in Italy and found that in Genoa and Florence the water was contaminated. Checking for hardness, calcium, sulphur, fluoride, heavy metals and traces of pollutants, the water samples in these two cities were found to contain dangerous levels of lead—three times over the accepted limit of 24.8, particularly at locations where traffic is heavy, and in Florence where there is a concentration of tourists in the Piazza della Signora. A representative of the Corporation of London expressed a similar concern about the water fountains on Hampstead Heath—a popular venue for walkers and runners, where drinking fountains offer a welcome site for the quenching of thirst. As he explained—'I would never drink from those fountains, they are dirty—dogs pee in them'. In 2017, the Rome city government also introduced regulations aimed at protecting its more elaborate fountains, banning picnicking tourists from some of the most famous fountains and raising the fines for those who take an illegal dip in their tempting waters. In 2017, two tourists were fined 900 euros for washing their feet in a fountain, while a local Roman was arrested for swimming nude in the Trevi fountain.

Istanbul

The emergence of a different set of relations of power materialized in water took place in another city noted for its fountains—Istanbul—in the eighteenth century. During that century there was a phenomenal proliferation of fountains: From 1703 to 1809, 365 fountains were built in the city. Hamedeh (2002: 123) suggests that the relentless construction of fountains can neither be attributed to improvements in the water supply, especially in the context of a corresponding decline in other water structures such as baths, nor in a population explosion at the time or a programme of repairs to urban infrastructure. Rather, she argues, this unprecedented fountain building can be accounted for by a broadening of the spectrum of elite groups who became building patrons, whose social diversity covered a broad continuum, from members of the central bureaucracy to the lower ranks of the military, craftsmen and artisans, with a notable growing proportion of women. Such a profile of fountain endowers represented a clear contrast with the dominance of the ruling elite in fountain construction in the sixteenth century (grand admirals and viziers), and the high-ranking military and palace agas (chief eunuchs of the imperial harem) during the seventeenth century (ibid.: 124).

Hamedeh suggests that the attraction of such investment lay in the confirmation of one's social status and power (ibid.: 124–5); these were affordable and small structures, where the name of the patron or a poetic inscription on the foundation were a visual representation of new and rising aspirations.

In Istanbul, Hamedeh (ibid.: 141) argues that their 'physical and institutional detachment from the religious institutions to which they had been connected' such as mosque walls or the inner city madrasa, and their new location in public squares, market places and public promenades, generated new forms of social interaction in the form of picnics, excursions and social gatherings. Quoting the chronicler of the city of Beykoz's descriptions of the effects of the new fountain built there, she describes the new vibrancy of the space: 'the place quickly turned into a summer-time recreational spot ….where people gathered to sit around the fountain or under its eaves…enjoying the sound of its gushing water…with the cheerful fountain the shore of Beykoz was brought to life' (ibid.: 142).

Paris

Paris is another European city where fountains in public and iconic spaces also have a long history, though not one that stretches back over the centuries to antiquity. Having researched the history of Dijon from the fifteenth to the nineteenth century Henry Darcy—the Chief Engineer of the Department of Cote d'Or (of Darcy's Law fame), concluded in 1856 that the construction of fountains was always a matter of concern to the residents, not merely motivated by the desire to embellish the city, but for several different goals: 'to flush out the great sewer that crosses the city, to provide clear water in place of unhealthy well water, to make water flow through all streets at scheduled times…to have an inexhaustible source of water nearby in case of fire, to construct public washhouses, and to make an ample supply of water available to both industrial establishments and private individuals' (Darcy 1856: 34). His strategy therefore, was to design and implement a complex hydrological system for the city to carry water 12.7 kilometres from the Rosoir spring through a covered aqueduct to reservoirs near the city, which then fed into a network of 28,000 metres of pressurized pipes delivering water to much of the city with no need for pumps, since it was driven by gravity.

As in Rome, fountains are dotted across the streets and squares of Paris enhancing public spaces and embellishing the cityscape. One intriguing form of fountain in the city are the Wallace fountains, which are usually painted in a striking bottle green colour. Their provenance is attributed to one Richard Wallace, the illegitimate son of Richard Seymour—Conway—the 4th Marquess of Hertford, who was brought up by his paternal grandmother in the sparkling cultured world of Paris during the Second Empire. Living through the political turmoil of 1848 and 1871, he finally left Paris to act as custodian for his father's art collection (Sykes 2003: 8). Already recognized with the highest civilian distinction (Commander of the Legion d'Honneur) for his charitable work in the city, his parting gift was to donate a million francs for the relief of the poor, and the offer of fifty drinking fountains to be designed manufactured and installed at his expense. In 1871, Paris was supplied water from springs transported by aqueducts, but only half of the dwellings had piped water; Wallace's gift was seen as of huge benefit to the poor for everyday domestic use, as well as providing cheap refreshment for those strolling along the Paris boulevards during their time off. An article in *The Times* October 1871 wrote (Sykes 2003: 16–17): 'to the paterfamilias who takes the whole of his family out with him to the Bois of a Sunday and to whom 30 cents per glass for a thirst compelling beverage is a serious consideration, to these persons and to a good many others, among whom may be comprised almost the whole of the female population of Paris, Sir Richard Wallace will appear in the light of a benefactor'. These ornate fountains, of which 82 adorn the streets of Paris today, are known by many Parisians as the Wallaces.

Assembling Publics: Fountains and Public Space

> The power of water it is so much about the basics of human life. When it is there and it is in your environment I think there is an immediate and instinctive reaction to it—and it doesn't really matter what form it is. …I think it makes you feel like you are safe… it is all about the emotions you can create in your audience….you can change the feeling of something… my favorite time is on completion going on ordinary day to see how ordinary people react to it—it I like it has always been there—I think that is the sense of what it is—normal has become nice—rather than a struggle—you can change a space and make it experienced very differently and make interaction between people different. (Victor Callister, Enhancement Planning Officer, Corporation of London)

In this next section, I turn to thinking through the power of water to assemble multiple publics in public space. The nature of public life and its particular blends of cultures, as ways of life and forms of aesthetic expression, are defining elements that differentiate cities and give each one its distinctive atmosphere; in this assemblage, I suggest fountains play a part. Before looking at how this plays out at one particular fountain, I want to take the reader on a short journey through some of the thinking on public space to make sense of why fountains might matter.

Publics and cultures, deeply entwined as I would argue they are, have complex, sometimes conflictual, relationships that define the politics of the urban, combining or drawing together different spatial scales and temporal trajectories into the particularities of place. If we take a look at fountains in city space we can begin to see how they involve the formation of multiple publics in an ongoing process of openings and closures of publicness that circulate and implicate subjectivities, bodies and materials in various ways. This brings out several contemporary themes in understanding publics and cultures—public as process, as multiple and involving forms of connection and communication that do not just involve humans (Bridge and Watson 2010). These themes include understanding the importance of affect and emotion alongside, or opposed to, rationality in public discourse. This means paying attention to bodies as well as minds in the constitution of publics. It also suggests how the public is constituted through materials as well as social action and discourse, which leads to questions about the role of buildings and urban infrastructure and services as part of the public realm.

The experience of being in public amongst strangers in the city has been seen to perpetuate urban divisions through forms of blasé behaviour and indifference. This is what Richard Sennett (2010) explains as the 'mask of rationality' that Georg Simmel (1948, [1903]) had previously identified as mutual indifference between urban dwellers in public to protect them from the unbearable nature of emotional contact and over stimulation in the city. These ideas of rationality in the city contrasted with arenas of political philosophy that saw the very idea of a public as the capacity to stand apart from one's interest and assumptions or social mores of the community. If we think about Hannah Arendt (1958) and Jurgen Habermas' (1984) arguments—they both had an idea of rationality as the ability to transcend interests and partial perspectives from community through rational discourse. Sennett (2010) puts forward the alternative approach to urban publics which sees the importance of the public in the city as performative rather than rational. This is the idea that differences can be overcome in public discussion and encounter through style of public address and bodily behaviour that cut across or disrupt social and cultural divisions. That is why Sennett (1970) stresses the edges or boundaries and the 'uses of disorder' between communities as the crucial space in which the public might be established (Sennett 1970). Sennett suggests that the capacity to dramatize difference through encounter in the spaces of the city often gets closed down by urban infrastructure and architecture that inserts boundaries rather than borders. He argues that buildings are over determined in their specifications and that architecture should plan for change and flexibility. Buildings should in this sense be incomplete, and technologies (such as time-based traffic restrictions) could help open up the spaces of encounter against the prevailing trend of the privatization of space through car use and loss of dedicated public space. On a related track, I have argued for the importance of more liminal and marginal spaces to open up the possibilities for rubbing along and encountering unknown and different others (Watson 2006).

Seeking to overcome the divisions between ideas of the rational and the performative in the constitution of the public in the city, others (e.g. Bridge 2005) have drawn on the work of American pragmatist philosopher John Dewey to suggest how performative bodies and rational discourse are interleaved as forms of communication in a continuum of human intelligent response to the environment. Dewey saw body-minds, emotion and cognitive reflection, as part of the same intelligent system of human organisms communicating to deal with the diverse impulses of their environment. These are forms of what Dewey called 'transactions' involving mediatised communication as well as face-to-face encounters, transaction implying that the subjects and objects involved were not rounded out and complete—but in process. This idea of the subject in process links to my argument for the public as process (Watson 2006), and to a wider idea of subjectivity beyond the subject formed through public discourse (Habermas 1984). The point Bridge makes is that this form of qualified impartiality can come about from the sheer diversity and intensity of impulses (including affect, bodies, emotion and 'thinking') that urban life affords. Public fountains can represent such a space of intense urban experience.

The diversity of impulses of the city includes its material and non-human environment. In relation to publics and cultures it involves how materials are cultural and act back on culture and how what Amin calls the 'situated surplus' of the material environment predisposes relations in public. This connects to the importance of the technological and the non-human in various 'assemblies' that constitute publics outside the realm of professional politics (such as laboratories, supermarkets) or over interests in the environment (a river): how things can make publics (Latour and Wiebel 2005). As overwhelmingly dense and intricate assemblages, cities are concentrations of the possibilities of new publics and new spaces of the public.

The importance of body as a form of disposition or habitus (to use Bourdieu's language) connects to three other themes. The first is the significance of the presentation of bodies involving body shape, clothes and styles of movement—what Goffman (1971) called body gloss. Bourdieu (1984) has noted how different types of gait and inclination of the head can express class position. How bodies present certain social/class dispositions is important in how people come into the co-present public and react to others. So much is communicated before a word is spoken and this has consequences for thinking about the public in terms of social interaction in the 'public' spaces of the city. Also important is body performance. Related to presentation, this is the way that bodies convey information and receive it from others. Sennett has shown the importance of this for public, or theatrical, forms of communication that act as a reference point for wider forms of public discussion beyond particular interests. But bodies can also be barriers to communication through the pre-discussion communicative content of their presentation (in various racialized, classed or gendered ways). Yet it is possible that these same bodies can break down divisions by forms of disarming conduct in spaces of the city that (unlike Sennett's example) are more mundane and are not specifically the spaces for public discussion or communication (Bridge and Watson 2010).

Another strand in the relationship between bodies and publics is emotion. Bodies are often conveyed as the repositories and most visible registers of emotion. Emotion has always been set against reason. If discussion in public requires the use of reason then emotions might be judged to have no place in publics. Simmel's 'mask of rationality' was a protection from the sheer emotional possibilities of encounter in the city. Sennett suggests how the emotions are important in body performance in the publics of the city, but he sets that against the idea of rationality and the public. Much of the discussion of the importance of emotion in sociology and geography takes a similar route of separating out emotion from rationality: emotion as an alternative or something that gets out from under the dictates of rationality, especially as those might discipline and control through the body (as Foucault argued).

A fourth element of bodies and publics is the way that bodies translate affect. If we take Anderson's definition of affect as the transpersonal capacity a body has to be affected, affect is decidedly non-rational but might be important for considerations of the public. Pile (1996), drawing on Freud's work from which the idea of affect emerges, discusses how cites are full of affect, of hauntings or feelings of unhomeliness or the uncanniness of being lost but finding yourself in the same place again and again. The mood of the city and the way the body senses that mood can be important in setting the tone for urban encounter and publics that may or may not result from that. Sense, as well as linguistic representation, might be significant for urban publics. Sounds of the city, for example, can register mood and affect—so can water.

This extended preamble on different ways of thinking about the public realm and public space provides the context for suggesting that fountains potentially offer a democratic public space of soft borders, performativity, lively materials and a space of engaged bodies, meaning and affect. The Princess Diana Memorial fountain offers a powerful illustration of many of these arguments, not least because of the affective responses to Diana symbolically as the People's Princess. Designed by an American landscape artist, Kathryn Gustafson, at a cost of GBP 3.6 million, the fountain is located in Hyde Park London just south of the Serpentine Lake, and was officially opened by the Queen on 6 July 2004 to a blaze of publicity. It was received with enthusiastic acclaim. For Glancey (2004), architectural critic at *The Guardian*, it was a 'near perfect metaphor for the life of Diana? Water ebbs, flows, gushes and chuckles round and around this prescribed oval course of beautifully cut granite slabs, before filtering out to a wider and more receptive world in the guise of the Serpentine with the tall and masculine buildings on its skyline receding into the summer haze. The cycle of a princess's life, you see, with all its ups and its downs, and their ultimate draining away'. The imperceptible edge where the water fades into its surrounds exemplifies Sennett's soft borders.

This artefact, more than many other contemporary UK public water feature, assembles complex relations of affect, identity, materiality and complex technologies and materials in a heady mix which disturbs both the idea of what public monuments are supposed to be on the one hand, and fountains on the other. Gustafson's conception of the fountain was that it was to reflect Diana's personality of inclusiveness and accessibility. During a BBC interview, she explained that the concept was based upon the qualities of the Princess that were the most loved and cherished. The sophistication of its design was enabled by state of the art computer guided cutting machines of the Northern Irish company S. McConnell and Sons to cut the 545 individual pieces of Cornish Granite into an oval shaped fountain, which takes the form of a large oval stream bed about 50 × 80 m and 3–6 m wide, that includes in its centre, and is surrounded by, lush grass. The streambed is very shallow to enable easy access for paddling and wading, as well as quiet contemplation of the water from the edge. A further engineering feat pumps the water from the top point of the oval down a gentle slope on either side over a variety of steps, curves and rills, causing gentle turbulence and ripples in the water, before flowing into the tranquil pool at the end. Gustafson's intention was thus to reflect the both calm and troubled life of Diana.

As with many public spaces (Watson 2006), it was health and safety discourse, which became the arena for the mobilization of disaffected voices and contestation. Shortly after its opening several people fell in the water and had to be taken to hospital as a result of fallen leaves and algae, which had formed on the surface of the granite, rendering the surface slippery. In response, the apotheosis of health and safety institutions— The Royal Society for the Prevention of Accidents was consulted on the crisis, with the head of water and leisure safety laying the blame on the surrounding grass which, though aesthetically pleasing, he suggested encouraged children to paddle thus bringing in mud and making it slippery. A fanfare of negative publicity in the British press ensued, reporting the disgust of Diana's friends with the memorial—Vivienne Parry: 'it is a half-hearted, damp squib that is, quite frankly, dissing Diana even in death.' and the 'national hand-wringing as the nation came to terms with its latest Great Endeavour Gone Wrong'. Critics were quick to draw comparisons with other publicly funded monuments and structures such as the GBP 800 million Millennium Dome, and the GBP 18 million Millennium Bridge across the Thames, which had similar teething problems causing widespread dismay.

The fountain was immediately closed. On its second reopening with roughened stone to reduce the slipperiness, and signs requesting people not to walk or run in the water, it was beset with further problems. As an article in the Daily Mail (2004) reported: 'Heavy rain at the weekend turned the fountain site, in London's Hyde Park, into a mud bath. A middle-aged woman, believed to be a Spanish tourist, was taken to hospital after slipping in the mud on Saturday, just a day after the fountain was reopened following a month-long safety review'. A Royal Parks spokesman who argued that the ground had been saturated by heavy rain contested the supposed lack of safety. He further disputed reports that the fountain had overflowed, since this was prevented by a remote-control system, which reduced water levels if there is a heavy downpour, thus preventing the fountain from flooding. In his view it was 'simply an unfortunate accident and a one-off'.

Its third reopening in May 2005 followed renovations including a new path and drainage and extra resistant turf costing about GBP 150,000, constituting a matter of concern to Westminster Council who feared that the original aesthetic had been threatened (Mail on Line 2006). Grumbles continued over the following months, with sporadic newspaper articles on the overspent budget, but over time the restrictions were lifted and people and children can walk through the water or sit on the grass nearby. The fountain now engages multiple publics throughout the year. In a series of observations and interviews conducted over one summer, the power of the fountain to engender passions and affect, as a space where bodies mingled with one another in fleeting, and not so fleeting encounters, was clear.

Responses articulated a complex mélange of nostalgic representations of Diana, notions of appropriate forms of memorialization, national identity and republicanism, as the comments below illustrate:

> We come often—very calming and relaxing—even when kiddies there—at the weekend—we come very often—if it was a stuffy old statue I wouldn't have seen it more than once…it's lovely-the kids—that's what she (Diana) did—really appreciate it—she would have loved it here, she would have been out all day. I come when I am hot—come in the winter—don't put my feet in then! Kids are grown up now. Bring others when they are with us. (3 white women from Dagenham)
>
> I didn't like it at first but now walking round it I have grown very fond of it. It's lovely to see kids running around—it is really nice. I thought it was a waste of money at first—just this morning I've grown really fond of it…When I read about it I thought it was a waste of time. I don't have children. Not my thing. Seeing that little boy and girl walking round—

2 Public Water Features: Assembling Publics, Enlivening Spaces… 29

they did the whole circumference of it—I thought they are not going to walk down the waterfall, but they did. They are having fun. …it works. I will come again. (Older white man from the Isle of Man)

The mythologies surrounding Diana as the people's princess who broke the mould of stuffy royalty removed from the populous, who engaged with people in need or marginalized through stigma, such as those infected with HIV, or victims of war as in Cambodia or just simply with ordinary people in the street, marked her as a princess who mobilized intense emotions of love and admiration among a huge diversity of people both in the UK and across the world. These feelings remain potent today, often articulated through an intimate discourse of almost knowing the princess personally, as was clear from interviews at the memorial fountain:

I know she would have liked it herself—funny thing to say. They drilled a special borehole to get the water out. It's natural the way they did it—in itself it's green as it were—it's a good memory of her. She would have liked the idea—you go in there you see families children old people—all generations—and that's what she represents in my mind. The tourists love it—Diana was very popular abroad and it reflects in the numbers of people who come to see it. I see lots of Americans lots of Japanese—everyone—Diana was well loved everywhere. And I look at her sons now and she would have been so proud of them—they have become like her haven't they?

We love all the elements, the water, the natural stone … it's lacking in flowers—there had been all the flowers at the funeral—so they should be there in the middle—we loved Diana—she was peaceful and humanitarian one of the people—the kind of person who would appreciate a fountain like this. (Danish daughter and mother visiting London)

Reflections on the fountain also extended to views on the royalty more generally, on a spectrum from royalist to republican comments:

I was watching a lovely programme about the Queen and her horses—and Clare Baldwin was so good so natural. The queen's fantastic—she's marvellous… I am a big royalist. I like British culture. I am proud of British, not in a racist way. I like what it represents in a good way. It's got a fairness to it. It's got freedom here-it's got reputation for being a fair country—that's why everyone wants to come here. It's so true. I don't like extremisms of any sort. That's a worry. But the middle ground is good. In Britain you have a chance to say what you want to say. And that's good. And sometimes you get the BNPs trying to hijack that it's not good. I am an old fashioned Brit. (Middle-aged white man)

It's beautiful—people need to be a bit more careful about messing it up. When kids play in it—sweet wrappers get dropped in—other than that, it is lovely. It's escaping, it's very serene, it's water by water, its very relaxing—good they did this for her—out of all the royals she was the only one I have time for—what she stood for—she was a people person she wasn't aloof—that is what they used to be—now Kate and Wills they are taking after her and the royals are taking on a whole new popularity. (Young black Londoner)

Not all comments were so positive—the park rangers thought the costs were excessively high and a waste of public money, while for others the connection of Diana to the fountain sparked a far more disturbing set of responses where a nostalgic and melancholic sense of an England lost and betrayed (Gilroy 2004) which in the context of globalization and mass in migration quickly shifted in this conversation into outright racism:

A   35-year-old park ranger: 'A concrete sewer—it's an awful thing—a big lump of concrete—it's a mindless thing—it would be better if it was a statue—I don't like it'.
B   Older park ranger: 'When it was first made it was supposed to represent all her life and da di da but when people were allowed to stand in and paddle in it—really and truly from the offset it was wrong because as soon as it gets water in it was going to go algae. It's fenced off now isn't it?'

The responses very quickly turned to the waste of money and their sense of their class being let down and forgotten in a Britain dominated by the rich and foreigners—a response which was manipulated and mobilized by the Brexiteers in the referendum of 2016.

B   Do you really think that the money they spent could have been better spent—it cost a lot of money—I believe the pumps alone were over a £100,000—that's why—they broke down after a day—I think it is a waste of money. It brings a lot of people in—a lot of people see it—but there is a lot of people out here in the world—in London around that could have done with the money—a lot of poor people a lot of older people which have gone through the war—had it very hard and still are having it hard now whereas a lot of people are getting it easier.

A This country is disgusting, it's done.

B These old boys and girls brought their kids up—they are struggling now.

A I can't get a mortgage—all you see is bloody Somalis—had enough of this country round the streets I live—came to work this morning—8 Poles came out of the park probably can afford a house cos there is 8 of them. Us people trying to get a house for ourselves and a little bit of England—it's finished, done we are gone. Hate the country.

At the end of the interview I pressed the two men on their views of the fountain as a monument to Diana. Here they referred to their personal connection to Diana, as rangers in the park where she lived, who greeted them often and for whom they had a personal affection. From their perspective an old fashioned public monument, statue or plaque, rather than a fountain would have been far more appropriate:

B I think there could have been something people look at and say that reminds me of Diana—but when you looked at that did that remind you of Diana? If you see that without any signs—would you have said that reminds me of Diana? No.

A I don't think it represents her person—it's just a concrete ring. If you didn't read what it says, you wouldn't know. It could be something better… just a place to wash your feet in the summer when you have had a walk.

Their sense of helplessness and disaffection is summed up in A's final comment: 'We are nobodies—doesn't matter what we think—we just work here'.

Of all the fountains I explored, this was the fountain that revealed the most complicated entanglements of affect, class and identity materialized in a public water feature. The symbolic power of Diana, and what she came to represent, evoked strong identifications with her memorial on the one hand or articulations of Britishness and racism on the other.

Other fountains similarly constructed with soft borders, which are now integral to many private development and regeneration projects as we see shortly, attract large numbers of people on a daily basis.

Enhancing Regeneration

For more than two decades water has been central to the regeneration and revitalization of cities. Crucial to the ethos of these strategies is the making of publics and socio-material relations where bodies and water spaces are co-constituted. Typically, water front redevelopment has turned old industrial buildings and docklands into residential and recreational areas for higher-income households seeking water views and access to valued amenities. Public water features have been similarly deployed to enhance cityscapes and to engender a sense of community and belonging (Strang 2004, 2009). Here, I focus on one such site—Bradford Mirror Pool. This water feature, I argue, illustrates how the entanglements of advanced computer technologies with public spaces together assemble new possibilities for social interactions and encounters. Like many other contemporary water features, it was designed, conceived and constructed by a state of the art company 'Fountain Workshop' based in Kent, which prides itself on combining innovative design and engineering with an appreciation of the importance of water as a device for creating vibrant and colourful public space. In these two sites the pools are shallow with soft boundaries to allow for easy access and water play, with computer controlled fountains, which are set to different programmes to provide shifting aesthetic effects, moods, possibilities and surprise.

As David Brace, its creator put it:

> The thing about water is it's the only form of environmental art that is dynamic—water really humanizes the space—take the water away and it is a big piece of paving.

Changing technologies which enable each of the fountains in the space to do different things at different times, from making mist in the morning for those walking to work, to dancing wave formations, to jets which follow people as they pass through the space has generated an entirely new sense of water as public space. As David Brace explained: In the early 80s when I started—people from councils would look through catalogues and say I'll have one of those and one of those—jets—people made fountain jets. They worked out what pipes and pumps they needed and the engineers' had them designed. This was a time when fountains were seen as ornaments and protected from the public. As a young engineer, he became interested in what could be done with water, and how to develop material spaces which enhanced sociality. This socio-technical connection was taken one step further with the development of an app to enable the passers-by to control and play with the water jets and flows themselves.

When designing a fountain, the company takes account of the specific demographics of the locality and how people across different ages and genders might want to use the space, from older people sitting watching the display to parents with children playing in the water. Fountain Workshop is also attuned to the different environments and cultural practices where the fountains are located. In Copenhagen the company placed 3000 fountains outside a corporate bank to soften the landscape around the building incorporating the local windy weather conditions so that walls of water rise and drop in different parts of the square depending on where the wind is, and even moving people around the space in different directions across the day.

What kinds of publics and space have been constituted in Bradford as a result of these strategies?

This was a city that had gone into steep decline following the deindustrialization of the 1970s and 1980s as the coal and wool industries went into steep decline. By the turn of the century regeneration policies were firmly on the political agenda, with the objective of revitalizing the then rather run down city. The key player behind the construction of the mirror pool in Bradford was the urban regeneration officer Shelagh O'Neill in 2003. Initially, through the preparation of a lottery grant application, and business plan which was highly rated but unsuccessful, and later through the council and its partnerships, she succeeded in bringing the project to fruition in 2011, countering fears that people would push shopping trolleys in or slip over. Drawing on cultural memories of Bradford (Broadford) as a centre on the river for the wool and brewery industries, the concept was to create a shallow pool of water in the centre of the city in front of the Victorian town hall to completely reorganize the space from a site bounded and cut off by traffic, to an open, flexible and accessible piazza of water, which could be crossed on boardwalks, or entered easily from the edge, echoing Sennett's (2010) notion of boundaried rather than bounded space making social encounters more possible.

The result is a stunning water space which is lively, enchanting and full of surprises—such as the intermittent high jet, the early morning mist, chase or shy settings for the fountains which engage with those passing by in serendipitous ways. Mindful of health and safety issues, particularly in the context of the Princess Diana fountain Shelagh O'Neill sought out materials (palfrey and granite) whose surface remains rough and non-slippery when wet. The entire site is accessible to those with disabilities, with different textures to mark different spaces for those who are visually impaired. Mobilizing the notion that responsible behaviour can be encouraged through zero policing, and making spaces attractive and fun (e.g. allowing young men to kick their footballs high into the jet fountain) rather than imposing rules and restrictions, she persuaded local law enforcement officers and agencies to adopt a non-interventionist approach. As a consequence, practices that in most city spaces are forbidden—sunbathing by the pool, splashing in the water, kicking balls and so on, are circumscribed only by notices which encourage mutual respect for the environment and others in signs placed unobtrusively around the perimeter. Reinforcing an ethic of care, rubbish and chewing gum are removed each day, which diminishes the very practices of discarding rubbish without thought. According to the two council caretakers and fountain operators this policy is a success.

Steve a British Asian Yorkshire man (the eyes and ears of the place according to Shelagh O'Neill) put it this way:

> when the kids are off school you see families on a Sunday morning or afternoon with picnics…they sit on the grass and the kids are playing all day in the pool area—it's very popular with the community—when really hot you looking at thousands (1500–2000) of people—they come to paddle. The behavior is monitored from a distance with CCTV and radio, if things get out of hand they do step in—positive reinforcement of good behavior instead of bossing people around—they try to reason with people depending on who is there at what time.

2 Public Water Features: Assembling Publics, Enlivening Spaces... 35

Observing the site revealed a social space of diversity and encounter of an easy and relaxed kind rarely found in a non-commercialized city centre.

Animating space and creating the possibility for multiple publics, has motivated not only public regeneration projects, but also commercial redevelopments, such as the extensive redevelopment of Kings Cross at Granary Square London (Photo 2.1), where the developers contracted Fountain Workshop to enhance the public spaces across the site. In both sites, this designer, has been involved in creating subtle illuminations which in Bradford play with public/private boundaries making the space feel like a room, and laser projections which like the fountains play and engage with the passers-by in a complex socio-technical assemblage of bodies, movement, acoustics and light.

Other regeneration strategies argue for fountains as a form of public art. The Corporation of London, for example in its planning documents prides itself on its 'long tradition of providing, maintaining and

Photo 2.1 Granary Square Kings Cross. Sophie Watson

encouraging publicly accessible fountains, statues and memorials. Today, more than ever, the importance of "Public Art" is recognised in terms of its contribution to the enhancement of the City's townscape and as a valuable source of enjoyment for all those who live or work in, or visit the City' (City of London n.d.). During the period of massive expansion of the City as a global financial centre and the accompanying extensive redevelopment, the Corporation of London sought to encourage private investors to donate or fund water features in their designs. In the Barbican Centre, for example, even the water-cooling system is in the form of a lake and a variety of fountains across the public space. Victor Callister (assistant director Environmental Enhancement) enthusiastically described a number of water features that had been constructed in the City of London—one in a private development at Broadgate where water cascades down the steps into a square from a deck above the railway lines where trains can be watched and where: 'People interact with it. Children paddle in it. But they do regulate it—they have private guards who ask people to move on—it is very much international banking area—for a city community it is right for that—it fits in with that—right for that—through this summer this space has come into its own'. The extent of exclusions, of any kind, was difficult to assess. He also described a fountain at the Royal Exchange where the water feature is set under a dense canopy of parasol plane trees which combine to cover a half dozen bubbling water features—'it is special and magical at night'. He made the further point that: 'It is very noticeable that those spaces that have water attract children and the presence of children there lightens the mood—in the City of London there are generally no children—if you add water—suddenly children are there—lightens feel of space… they bring their costumes and they play after school.'

In another instance in Fleet Street, a fountain has been enrolled in a project with Goldman Sachs into counter-terrorist initiatives, where water cascades over stone blocks that deter entrance, evoking the fortress city notion explored by Davis (1990) in *City of Quartz.*

In the US, the concept of the swimmable urban fountain has been introduced in urban design initiatives across several cities, enabled by new sanitation and digital technologies, many of which reclaim and reuse water. In Los Angeles, for example, a dilapidated fountain in Grand Park was revitalized to create a new public space: the Arthur J Will Memorial Fountain has hosted performances by dancers in wetsuits, where a neon light show plays at night, and where children play on hot days. At the Crown Fountain in Chicago's Millennium Park, an interactive sculpture incorporating two giant LED screens inside glass-bricked monoliths stand on opposite sides of a wide plaza. Evoking ancient fountain motifs of deities and mythological figures, the fountains display images of 1000 different Chicagoans, who smile at first, then pucker their lips to reveal a stream of water that 'spits' into the granite pool.

City Branding and Place Making

Many cities across the world have constructed dramatic fountains to brand their cities for investment and tourism and to assert their presence on the world stage. The Fountain in the United Arab Emirates which opened in May 2009 represents the most ostentatious and extravagant—and wasteful of energy and water—fountain constructed to enhance a city's status as a centre of global capital. Built at a cost of AED 800 million (USD 218 million), it is illuminated by 6600 lights and 25 coloured projectors, it is 275 m (902 ft.) long and shoots water 240 feet into the air accompanied by a range of classical to contemporary Arabic and world music. The choreographed fountain system designed by a California based company, WET design, was built on a 30-acre man-made lake, at the centre of the downtown, with the force to spray 22,000 gallons (83,000 litres) of water in the air at any moment. More than 6600 lights and 25 colour projectors have been installed. The deep interconnections of the city and its symbolically powerful fountain with global capital is illustrated by DHL Express' 'most ambitious advertising campaign to date' when the global logistics company premiered an eye-catching light show in Downtown Dubai on 20 October 2011, which was choreographed in dramatic style to a specially reworked version of the Motown classic 'Ain't No Mountain High Enough'. The song is also the soundtrack to DHL's advertising campaign that was aired across the Middle East region. The fountain is animated with performances set to light and music and is visible from every point on the lake promenade and from many neighbouring structures. Reinforcing its symbolic role as a site of national identity, the first show of the day is the 'Sama Dubai', a tribute to Dubai's ruler Sheikh Mohammed.

In the US, Kansas makes similar claims to fame, marketing itself as the city of fountains. In the 1960s, the city government regenerated the central square using a fountain as the key feature. In total there are 49 jets of water that are propelled into the air up to a height of 60 feet—the unique element to this fountain is that there is no basin, rather the water is gathered from an underground chamber from which 2200 gallons of water per minute are propelled into the air. The fountain is illuminated by an array of coloured lights which shine through the water beams. There are many other such initiatives across the globe.

Thirst Quenching

There is something curiously democratic about the provision of water fountains in city spaces, where people share the source one after another, as they lean in to imbibe this basic force of life (Photo 2.2). Drinking fountains have a long and noble history (see also Darcy 1856). Following an investigation which revealed that no drinking fountains across the UK had been certified as drinkable, in 2010, the Corporation of London inaugurated a scheme to introduce new drinking water fountains within the open spaces and public realm of the City and to restore the City's historic drinking fountains. One rationale for its introduction was the City's commitment towards reducing plastic waste by providing alternatives to bottled water and towards creating a more pleasant and enjoyable pedestrian environment (Photo 2.2).

Victor Callister reported that in 2013 an investigation across the country revealed that there were no properly certified drinking fountains. According to Callister, water authorities were connecting to fountains that were not suitable for drinking, where fountains sit amongst trees with dogs and pigeons lapping at them. The city determined to tackle this by rolling out certified drinking fountains across the city with the result that seven were installed by the end of 2013 including a fountain

Photo 2.2 A drinking fountain in Venice. Sophie Watson

11 metres tall at the Guildhall—'a monument of Victorian philanthropy'—that originally had a cup on the side for drinking. Mindful of health and safety concerns, the fountains were constructed to enable water bottles to be filled rather than encouraging people to drink from the source to, supposedly, avoid contamination. Initially the fountains were tested every three months.

Conclusion

This chapter has explored the myriad of ways in which fountains in the city animate public space, assemble multiple publics, are deployed for urban regeneration and place making and branding and exist as symbolic sites of meaning. Fountains not only reveal some of the complexities of urban life and its textures, they also offer a window into understanding contemporary knowledges of hydrological technologies and design at different historical moments. Fountains connect bodies with water in unpredictable and serendipitous ways. An exploration of fountains in the city, as we have seen, also offers a lens into the political, socio-cultural, symbolic and spiritual life of different cities at different times, sometimes constructed to consolidate political power, at others to honour specific local or national figures, and at others for more functional purposes such as to quench thirst or enable bathing. Like many other often unnoticed objects in the city, located on a street corner, or in a square or municipal park, that are passed daily attracting little interest or comment, these sometimes beautiful, sometimes indifferent, material structures have a far richer and more textured story to tell that many city residents never get to hear.

References

Arendt, Hannah. 1958. *The Human Condition*. University of Chicago Press.
Barber, T. 2003. *Waterfalls and Fountains*. Neptune City, NJ: TFH.
Bourdieu, Pierre. 1984. *Distinction: A Social Critique of the Judgement of Taste*. Cambridge: Polity.
Bridge, Gary. 2005. *Reason in the City of Difference: Pragmatism, Communicative Action and Contemporary Urbanism*. London and New York: Routledge.
Bridge, Gary, and Sophie Watson, eds. 2010. *The New Blackwell Companion to the City*. Oxford: Wiley Blackwell.
Corporation of City of London. n.d. *Fountains in the City of London*. Corporation of City of London, Department of Planning.
Daily Mail. 2004. Diana's Fountain Turns into Mudbath, August 23, 2004. http://www.dailymail.co.uk/news/article-315123/Dianas-fountain-turns-mudbath.html.
Darcy, Henry. 1856. *Public Fountains of the City of Dijon*. Dubuque, IA: Kendall Hunt Publishing Co.
Davis, Mike. 1990. *City of Quartz*. London: Verso.
Edwards, Catherine. 2017. Rome Turns Off Its Public Drinking Fountains to Cope with Drought. *The Local.com*, June 30.
Forbes, S. Russell. 1899. *The Aqueducts, Fountains and Springs of Ancient Rome*. London and New York: T. Nelson.
Gilroy, Paul. 2004. *Postcolonial Melancholia*. Columbia University Press.
Glancey, Jonathan. 2004. Diana Memorial Fountain Completed. *The Guardian*, June 30.

Goffman, Erving. 1971. *Relations in Public: Microstudies of the Public Order*. Basic Books.

Guardian. 2017. *Vatican Turns Off Historic Fountains Amid Rome Drought*.

Habermas, Jurgen. 1984. *Moral Consciousness and Communicative Action*. Cambridge, MA: MIT Press.

Hamedeh, Shirine. 2002. Splash and Spectacle: The Obsession with Fountains in the Eighteenth Century Istanbul. *Muqarnas* 19: 923–948.

Latour, Bruno, and Peter Wiebel. 2005. *Making Things Public: Atmospheres of Democracy*. Cambridge, MA: MIT Press.

Mail on Line. 2006. Timeline: The Troubled Waters of Diana's Fountain, March 21, 2006.

Morton, H.V. 1970. *The Fountains of Rome*. London: Macmillan.

Pile, Steve. 1996. *The Body and the City*. London: Routledge.

Sennett, Richard. 1970. *The Uses of Disorder: Personal Identity and City Life*. London: Penguin Random House.

———. 2010. The Public Realm. In *The New Blackwell Companion to the City*, ed. G. Bridge and S. Watson. Oxford: Wiley Blackwell.

Simmel, Georg. 1948. *The Metropolis and Mental Life*. Translated by Edward Shils. Social Sciences III Selections and Selected Readings, vol. 2. Chicago: University of Chicago Press.

Squires, Nick. 2017. Rome Turns Off Its Historic 'Big Nose' Drinking Fountains as Drought Grips Italy. *Daily Telegraph*, July.

Strang, Veronica. 2004. *The Meaning of Water*. Oxford: Berg.

———. 2009. *Gardening the World: Agency, Identity, and the Ownership of Water*. Oxford and New York: Berghahn Publishers.

Sykes, Alan. 2003. *The Wallace Fountains of Paris*. AbeBooks.

Venturi, F., and M. Sanfilippo. 1996. *Fountains of Rome*. Vendome Press.

Watson, Sophie. 2006. *City Publics: The (dis) Enchantments of Urban Encounters*. London: Routledge.

3

Consuming Water: Habits, Rituals and State Interventions

Without water, there is no possibility of life. Humans need water to drink, to wash, to cook, to clean, to grow crops and for a multitude of industrial purposes from food to energy production. But how much water individuals and societies actually need is culturally contingent, historically specific and open to contestation. In many countries, water is taken for granted in a myriad of ways. Yet increasingly as climate change impacts on rainfall, the ice cap, sea levels and rivers, water can no longer be taken for granted even in places where access to water once seemed ubiquitous. In 2017, it took 3781 litres of water to produce one pair of Levi 501s and 16,000 litres of water were needed to produce one kilo of meat—a reason many vegetarians cite for their choice of diet. Some water activists have devised the notion of the water footprint to highlight the total volume of fresh water used in the making of products such as food, energy and clothing. Individuals thus also have their own personal water footprint that can be shifted, not just by changing everyday practices of washing and cleaning, but by changing the food they eat or the clothes they wear. Access to water, or the lack of it, represents one of the greatest forms of division and inequality across the globe—if not the greatest. According to the World Bank more than a billion people live in water-scarce regions and as many as 3.5 billion could face water scarcity by 2025. Arguably, water shortage will be the cause of the most severe forms of conflict over the next century, and generate migration at unprecedented levels as populations without access to water are forced to leave their homelands in search of a viable life. The significance of water to everyday life cannot be underestimated.

Water consumption is embedded in a complex assemblage of sociotechnical connections, which are constituted differently over time and space. No one water system or form of provision is the same. Thomson (1877: 95–7) writing of a moment in Victorian street life, describes the displacement of the old wooden water cart by a perfectly constructed iron tank fixed on the framework of a specially fitted cart. Not only did the tank contain a large amount of water, but also it was not prone to leakage and enabled the water to be discharged with some force thus facilitating road cleaning. This is but one moment in a long history of changing technologies, and their interconnections with social relations, which have taken different forms across the world, and which have developed at a different pace, such that in many countries water is still only available to the majority of the population from a pump in the street.

Water consumption is embedded not simply in a complexity of sociotechnical assemblages, it is also produced within an intricate pattern of unremarkable habits, practices, rituals and conventions which differ from place to place. Shove (2003) suggests that there are three domains of daily life—comfort, cleanliness and convenience, which are subject to recent and radical change, which have particular meanings at particular times, with consequences for behaviour, and which nevertheless fall into the realm of the taken for granted. As Taylor and Trentmann (2011) point out, things and practices are never homogeneous, but always are assembled out of a multiplicity of traditional and modern forms. As new technologies develop, so new habits and sensibilities evolve but not in a smooth symbiotic fashion: piped hot water made possible new private routines—while also leading to political mobilization as middle-class rate payers demanded their rights (Taylor and Trentmann 2011). Thus, new technologies generate new everyday practices and politics and a changing relationship between the boundaries of private and public life. In this sense, thinking about the consumption of water implies paying greater attention to the changing materialities and infrastructures of public life in any one place at a particular moment. No place is ever the same and the fabric of everyday life is in a constant state of flux.

Patterns of Consumption

Three-quarters of the earth's surface is covered with water, yet 98% is salt water and not fit for consumption. The disparities in water consumption between different parts of the globe offer a stark illustration of the shameful inequalities between the richest and the poorest nations. In 2006, 3.9 trillion gallons of water were consumed in the US per month. The average American uses 800 litres per day, with greater use in the Western states such as Arizona and Utah, where large amounts of water are used to maintain garden lawns and flower beds. Typically, water consumption in many European countries is lower than in the US; in Germany for example, the water consumption per person amounts to 121 litres water per day of which about 1/3 is available for toilet-flushing, 1/3 for body hygiene and another third for laundering, washing the dishes, cooking and drinking. The profligacy of water use in the richer nations is no more starkly illustrated than in the estimated 2.5 billion gallons of water used each day on irrigating the world's golf courses.

By contrast, in Africa about 85% of the water is used in agriculture, while only 10% is used in households and only 5% in the industry sector. Many people have to survive on just 20 litres per day—the equivalent of one shower of 1.5 minutes taken in the US. According to the UN, a human being needs 50 litres of water per day in order to prepare meals and to have enough for personal hygiene. The estimated amount of water consumed by 60,000 villagers in Thailand, on average, per day is 6500 cubic metres—the same quantity of water used in one golf course in Thailand on a single day. According to the World Health Organization, an estimated 2.4 billion people lack adequate sanitation and 1.2 billion people (one fifth of the world's population) live in areas of water scarcity and a further 1.6 billion people face economic water shortage—that is where countries lack the infrastructure to take water from rivers and aquifers (WHO/UNICEF 2018). Ninety per cent of wastewater in developing countries is discharged into rivers and streams without any treatment. The average distance that women in developing countries walk to collect water per day is 6.4 kilometres and the average weight that women carry on their heads is 20 kilos. Women in sub-Saharan Africa spend

50–80% of their time fetching water. Water scarcity is the reality of daily life for people in many parts of the world, with stark figures for the proportion of the population affected by water scarcity for at least one month a year being 96% in India and 83% in Pakistan and over 60% in Australia and Turkey. According to the World Bank, by 2025, as many as 3.5 billion people could face water scarcity. The World Economic Forum, every year from 2014 and 2017, has ranked water crises in the top global risks.

Not only do people across the world have highly different water consumption rates, the amount they pay for water is at stark variance, with once again the poorer nations paying the highest costs, relative to income—where the majority of the population is entirely reliant on deliveries of water in large containers or bottles or one tap for many households. The price people pay for water is determined by three factors: the cost of transport from the water source to the user, total national demand and price subsidies. The treatment to remove contaminants from the water can also add to its cost. The price of water is rising globally according to Earth Policy Institute's latest report. Over the past five years, municipal water rates have increased by an average of 27% in the US, 45% in Australia, 50% in South Africa and 58% in Canada. In poorer countries, the increase is far greater, with a fourfold increase in Tunisia, for example, in the price of irrigation water. If like for like costs of water are compared between the US and Guatemala, the cost of water bought from a vendor in Guatemala will be at least 5 times greater than in the US.

Corruption represents a further difficulty, where wealthier residents fuel a black market in water. In New Delhi for example, during the hot season, an estimated 2000 illegal tankers drive around the city selling water to the poorer inhabitants derived illegally from the city's ground water and often polluted, leading to outbreaks of disease. Another route to profit is the reselling of water, which in Lesotho is estimated to represent the water supply of 31% of the population. These water mafias—according to the *Cities without Water* report—pay off city officials and the police who collude in the illegalities and reinforce the vulnerability of the poor.

Leakage and Infrastructure

Water is not always purposefully consumed. To the contrary, large quantities are wasted through poor infrastructures and leakages, through pollution or simple mismanagement. In June 2017, the environment editor of the Guardian (Carrington 2017) drew attention to the vast amount of water in the UK that water companies are losing through leakage—an estimated 20%, provoking consumers not only to constant complaint, but also contributing to the resistance of customers to reduce their own consumption in the context of what they see as the profligate wastage of the providers. According to the water industry regulator Ofwat, more than three billion litres of water leak out every day at a level unchanged for at least four years and just 7% lower than the level in 2000 (Carrington 2017). Not only have some of the major companies across the south and east seen no significant reduction, one company, Essex and Suffolk Water has seen leaks rise by 15%. Thames Water is the largest water company in England and also the leakiest, with 20,500 litres escaping every day per kilometre of pipes, more than double the national average, and equating to 171 litres per property per day. In New York City, 30% of the water is estimated to be leaking (Murley quoted in Anand 2015: 38). It is interesting to consider this widespread leakage in contrast to the objective reality presented by water management companies in the form of bills, annual reports and projections of profit. As I have argued earlier, this illustrates starkly the uncontainable nature of water as a substance.

Faulty infrastructure is an even greater problem in some of the poorer cities of the world. In Mumbai, where the century old network of fragile pipes are crumbling, the water system is already at crisis point, with about 20% of all the city's piped water escaping into the ground through leaks, and periodic ruptures in the mains which contributes to the sense of a crucial urban resource being wasted before reaching the people who need it. All this means that Mumbai, with a notional per capita water availability of 180 litres per person per day—30 more than in London, fails to deliver a regular supply of water even to its wealthier citizens (McFarlane 2013: 121).

Water is not only wasted through leaks, it is also often not drinkable and even when water is available is likely to be polluted—1.6 million deaths are attributed to dirty water and poor sanitation. A report released in Karachi in July 2017 (Sahoutara 2017) analysing surface and underground water sources found that up to 90% of water supplied in the city was unfit for human consumption due to the presence of bacterial contamination. Faecal contamination was detected in 40 of the 118 samples taken. The report concluded that the main reasons for the pollution were the supply of raw [untreated] water with filtered water, improper chlorination, mixing of sewage water and water supply lines/sources and the silting of pumping stations/water storage tanks (Sahoutara 2017). Similar stories of contaminated water are told across the world, for example, 35 of the 40 water samples collected in Hyderabad were recently found to be unsafe for human consumption. This is the context for the proliferation of bottled water in many cities of the world. Even the imagined idylls of rural England have not escaped the occasional case of polluted water. In 2014, following a case brought by the national watchdog, the Drinking Water Inspectorate, Coventry magistrates fined Severn Trent Water GBP 66,000 after the utility pleaded guilty to 11 counts of supplying through the taps in the village of Broadway Worcestershire that was water 'unfit for human consumption'.

Salt in the water can be another problem. In 2015, in Nairobi, tests on a vast aquifer found in Kenya's drought-wracked Turkana region revealed that the water was too salty to drink (Ruvaga 2015), dashing the hopes that had arisen with the 2013 discovery of underground lakes in this arid region where some 135,000 people needed water and food assistance. After drilling 350 metres underground, saline levels were found to be seven times higher than the level considered safe by the World Health Organization (WHO). A local water officer contested the conclusion and suggested that there were some adjoining wells where the saline limits were acceptable which would have alleviated the problem. Nevertheless, this was a dire finding in a region susceptible to recurrent drought which decimates livestock kept by the traditional nomadic herders, and where armed conflict over limited resources is common.

Bottled Water

One of the most pernicious forms of water consumption is enacted through the promotion, production and distribution of bottled water—the vast majority of which is in plastic bottles. In some countries such as India where the water system fails to provide clean and regular water to many of its citizens, high levels of bottled water consumption are no surprise. In countries where fresh and clean tap water is readily available through taps, there is little justification—or indeed rationale—for drinking water that is both expensive and reliant on containers that have detrimental environmental effects. An estimated 100 million bottles are used worldwide every day with considerable variation between countries reflecting economic, political, technological and cultural differences. According to the World Atlas (http://www.worldatlas.com/articles/top-bottled-water-consuming-countries.html) in India 10.4 billion gallons of bottled water are consumed annually. In France, there is a long tradition of drinking bottled water, with half the population consuming bottled water on at least one occasion every day (2.41 billion gallons in total). In Germany, the widespread availability of bottled water products in retail stores, as well as vending machines, has led to their increasing consumption. The increased consumption of bottled water in recent years (3.17 billion gallons) is attributed to concerns over the quality of the country's tap water—an issue which is even more salient in Thailand where the deteriorating state of the country's ageing plumbing systems has led to an annual consumption rate of 3.99 billion gallons. In Brazil, the rationale for the 4.8 billion gallons consumed annually is the high number of people living in the urban slums where there is a limited fresh-water supply and poor water purification practices that have put thousands of people at risk for contracting water-borne diseases. A similar level of consumption is evident in Indonesia which is attributed to rapid urbanization and a growing population rate, where many local Indonesians are becoming better educated, and thus more concerned about the quality of their tap water (http://www.worldatlas.com/articles/top-bottled-water-consuming-countries.html).

The socio-technical and cultural landscape in each country is thus different. Water bottles, as Hawkins (2011a, b) and Hawkins et al. (2015) suggest become participants in assembling particular relations between water and life that do not automatically overrule other water regimes but that interact with them in complex and unpredictable ways. Moreover, the plastic water bottle has the capacity to 'discipline water's biophysical unruliness; to appear to manage risk and scarcity; and to individualize provision' (Hawkins 2011b). Bodies are implicated centrally here particularly in the sense that an increasing emphasis on regimes of health and fitness has mobilized desires for endless rehydration which are easily met by the plastic water bottle.

Bottled water is frequently promoted as a healthy alternative to tap water and marketed with images aiming to evoke the 'natural' environment, such as bubbling springs or mountain ranges. The reality of bottled water is far from natural or environmentally sound. Water in bottles is often simply purified tap water rather than extracted from springs and streams. Equally contradictory is the fact that the very bottles in which water is packaged represent an environmental and health disaster, and the plastic used in bottles has been associated with health problems such as cancer and diabetes. Poisoning through diet is another consequence of plastics, since seafood is also affected as fish and shellfish contain toxic chemicals at concentrations as high as nine million times those found in the water in which they swim. The effects on energy consumption represent another negative impact since more water is used in the production of bottled water, than is drunk in any one bottle (an approximate ratio of 3:1) and 17 million barrels of oil are needed to produce plastic bottles yearly. This could fuel 1 million cars for one year (https://www.banthebottle.net/articles/10-startling-facts-about-bottled-water).

The absence of everyday practices of recycling represents the most serious problem with typical estimates of only one out of five plastic bottles being recycled. The material from which they are manufactured is petroleum based and thus takes 700 years to begin to decompose, since the bacteria that typically break down organic materials are unable to break down petroleum-based plastics. Plastic bottles end up

3 Consuming Water: Habits, Rituals and State Interventions 51

in landfills across the countryside or in the ocean—one calculation is that at least 10% of the 100 million tons of plastic we use every year end up in the oceans which is equivalent to the weight of 700 billion plastic bottles. As a result of the action of ocean currents, there are 'plastic soups' around the world. The exact size of these is difficult to accurately assess but *The World Counts* website suggests they could cover up to 16 million square kilometres, or the combined size of Europe, India and Mexico (http://www.theworldcounts.com/counters/ocean_ecosystem_facts/plastic_in_the_ocean_facts).

Predictions are that if nothing changes, by 2025 there will be one ton of plastic for every three tons of fish in the ocean (*Oneless*).

Water bottles are devices which constitute a diversity of political action, contestation and controversies calling new and concerned publics into being and creating new connections and networks (Marres 2012). Bottled water has generated numerous political situations, and as Hawkins et al. (2015: xiii–xix) argue, bottled water is not simply a new development in the beverage sector, it also brings into being new and surprising relations between and among plastic water bottles, water and the drinking body. This resistance across the globe to plastic water bottles is illustrated in a growing number of campaigns and websites seeking to reduce their consumption. Many such initiatives advocate drinkers to buy a reusable water bottle or donate such a bottle to encourage participation. *Oneless* in London is one such campaign which deploys social media tactics to urge supporters to take a 'selfie' with their favourite refillable water bottle and to tweet it. They have also supported the Zoological Society of London's decision to cease selling single-use plastic water bottles across both of its zoos and have partnered with Marine CoLaboration who are working to make London the first capital city to stop using single—use water bottles, and with Selfridge's to remove all single-use plastic water bottles from their food halls and restaurants. The *London on Tap* campaign—which involves collaboration between Thames Water and the Mayor of London similarly promotes the adoption of tap water in London's restaurants, bars and hotels and encourages customers in bars and restaurants to ask for tap water rather than feeling obliged to drink water from expensive bottled brands.

Strategies for Water Consumption

There are two countervailing patterns of water consumption. On the one hand, many cities of the world are beset by problems of water shortages, limited or no access to clean water and excessive costs. Consumption is characterized by shortage. On the other hand, the citizens of cities in the more affluent and developed parts of the world typically have access to clean and affordable water at the turn of the tap, and take its continued availability for granted, even in circumstances where there is increasing water stress. There are thus uneven and complex issues to be addressed across the globe, with strategies for increasing water distribution and provision needed in some parts of the world, and strategies to reduce water consumption needed in other parts of the world. Such spatial differences are implicated in relations of power, where poorer nations lose out. These are multifaceted questions which warrant books in their own right. For my purposes here, only several directions of policy can be briefly indicated.

African cities provide a stark illustration of places in severe water stress as we saw earlier. Many of the problems seem intractable with corruption of the water mafias representing one hard nut to crack, and failing infrastructure another. Population growth in many urban centres, for example, sub-Saharan Africa's urban population is predicted to double over the next 20 years, which implies a worsening of the problem of an unimaginable kind, as more and more people need clean and affordable water. Nevertheless, some cities are embarking on strategies that will have positive effects. One route to increasing water provision is desalination, but this process is both wasteful in terms of energy and expensive. Water stress is not confined only to the poorer parts of the world. Cities in California, for example, are constantly subject to water restrictions, often resulting from profligate use of the available water supplies.

Countries where the availability of water has to a varying extent been taken for granted, in recent years have introduced new strategies to reduce water consumption in the recognition that growing populations, or the effects of climate change, necessitates a more proactive approach. Water activists argue for multiple methods of communication including interventions in social media, outreach to children through the school curriculum and public meetings with stakeholders. The separation and treatment of grey water, storm water and rainwater is another strategy, including recycling—though there is considerable resistance in some parts of the world to the idea of recycled water, particularly from sewerage, which is met with disgust. Integrated water management bringing together departments, agencies and stakeholders, is another route to sustainability. Notably Australia has been at the forefront of water saving strategies and technologies, in the context of an environment persistently threatened by drought and water shortages. From 1995, Sydney Water set in train some of the largest water efficiency programmes in the world, mobilizing their customers to actively participate and to observe water restrictions and *Water Wise Rules*. Recycling of grey water is now common. It is from this country that some of the most insightful critical analysis has also emerged (Allon and Sofoulis 2006; Randolph and Troy 2008; Head 2008; Hawkins 2011a, b; Sofoulis 2005, 2011, 2013; Head and Muir 2007; Maller and Strengers 2013).

Pricing strategies also have an effect. In the UK, there has been a long tradition of flat charges for water across many parts of the UK which has led to an ignorance of, and indifference to, the cost of household water consumption. A report commissioned by the Department for Environment, Food and Rural Affairs (Walker 2009) found most people considered that water remained cheap and plentiful. Some water companies charged a percentage of the property's rateable value, which varied between Water Companies; others used an assessed volume charge based on the size and type of the property or the number of occupants. Households could also request a water meter, though the minority pursued this strategy, even if it would have benefited them to do so. In the context of a climate where rainfall is often high, a dominant imaginary was of water abundance, leading to cultural practices of extensive water consumption for domestic purposes, despite regular bans on garden watering during hot and dry summers. Other environmental policies may also have knock on effects on water consumption. For example, though water bottles might be seen as the individualisation of a collective problem, the effect of carbon tax could be to make disposable plastic prohibitively expensive as petroleum extraction itself becomes less profitable.

In recent years with the growing digitalisation of everyday life, the introduction of smart meters as energy saving devices has become increasingly prevalent. These smart initiatives are increasingly entering into domestic life and 'becoming a field of social control that makes intrusion in a person's private life quite natural' (Vanolo 2014: 894). Smart initiatives to reduce water consumption and shift everyday cultural practices represent a particularly interesting initiative since they enter the commonly understood to be private sphere of the home, disrupting everyday routines and challenging embodied and cultural habits and practices in unusual ways. Many of the smart visions operate with a notion of the 'resource man' who will make rational decisions on the part of his household to adjust everyday practices as advised and adopt the new smart devices (Strengers 2013). Lying behind such a figure is an educated, able, knowledgeable and technologically savvy subject who knowingly participates in the intervention. Thus, there are inherent assumptions about individual behaviours and cultural practices that go hand-in-hand with the introduction of smart meters, which I suggest disrupts their effectiveness in controlling the demand side of water resource management.

To explore this contention, in the rest of this chapter I chart an intervention by Britain's largest water company Thames Water to reduce water household water consumption, in order to highlight the complexity of cultural practices, subjectivities, dispositions and affective responses of its customers when faced with strategies to shift their attitudes and practices in relation to water, which had been revealed as significant elsewhere. For example, in their study of everyday community attitudes and practices around water in Western Sydney Allon and Sofoulis (2006) found that water and water saving devices produced a diversity of domestic routines, practices and habits, which derived from multiple human-cultural-natural-technological relations that produce and reproduce socio-cultural differences in a complex assemblage of factors that are distributed across space, time and bodies. Other Australian research (Jenkins and Pericli 2014) also exposed a diversity in domestic uses of water across activities such as showering, bathing, washing, cooking, toilets and so on where class (defined through income and education), age and gender have also been central to explaining why some water consumers are willing to change their behaviour and some are not. In another study, Randolph and Troy (2008) revealed the

3 Consuming Water: Habits, Rituals and State Interventions

significance of housing tenure and internal domestic architectures on perception of water use. Robinson et al. (2014) reported how the warmth of a bathroom affects the length of time people spend in the shower.

The story begins with the publication of a report from the Department for Environment, Food and Rural Affairs, which concluded that there was a growing trend for increased demands for water, pressure on water supply—particularly in the South East—and climate change projections of drier summers and seasonal flooding, water authorities across the UK from 2010 were propelled into a range of strategies to cut water consumption. Consultations for the report (Walker 2009) suggested that a charging system could incentivise a more efficient use of water to ensure water supply, and that the introduction of metering represented the fairest approach to charging which could be beneficial particularly to those on low incomes for whom affordability was a key issue. A further anticipated advantage of the installation of meters was that they allowed household demand to be more closely matched to supply, through the possibility of water companies collecting more up-to-date information on where water shortages are occurring, enabling appropriate and timely advice to households and thus avoiding the introduction of the more extreme types of rationing.

In 2010, Anglian Water which was operating in an area with an average annual rainfall of a third less than the rest of England, introduced the 'Love Every Drop' campaign to put 'water at the heart of a whole new way of living' promoted by its chief executive—who having travelled the world knew first-hand 'how essential water is to life, and what it means for people and the environment to be without it'. Simultaneously, Southern Water initiated its programme to install meters across Kent, Sussex, Hampshire and the Isle of Wight. Thames Water followed with its smart metering programme, aiming to install smart meters in 3.3 million properties by 2020. Smart meters are a regulatory device to reduce domestic water use which are connected to a wireless network, so customers can view their water consumption online around the clock, gaining greater control of their usage and their bill. In each of these initiatives authorities mobilized discourses of personal responsibility—'doing your bit' and a greater understanding of water's role in carbon emissions: 'Where there's water, there's carbon—and quite a bit of it. Every bath, flush or glug has CO_2 built into it, thanks to all the processes it takes to get it to the tap. So using less water is good news for the planet' (Anglian water), in their attempts to shift an embedded understanding that water will just keep on flowing. Alongside regional interventions there has been a host of other government and voluntary sector initiatives which have enrolled water as a more prominent public matter of concern—including the Energy Savings Trust 'At Home With Water', with the objective of educating publics, also through a device to calculate household average use—the Water Energy Calculator, and Waterwise's 'water saving weeks'.

Based on research conducted in Bexley and Greenwich on the Thames Water strategy to roll out smart meters, I explored the intervention through three frames: affect, habit and the meaning of home in constituting specific individual and cultural responses to the intervention. The notion of affect here draws upon Anderson's (2009: 78) attention to collective affects that 'press upon' life and the idea of 'affective atmospheres: serene, homely, strange, stimulating, holy, melancholic, uplifting, depressing, pleasant, moving, inviting, erotic, collegial, open…' which draw attention to the often-unspoken responses to interventions imposed by others. Habit is another useful lens for focusing our attention on the unnoticed responses of individuals in everyday life. The meaning of home foregrounds a third terrain, again sometimes unnoticed, but central to highlighting how everyday practices emerge from understandings of private/public distinctions, which are culturally and historically specific. I want to argue that these are further mediated, and co-produced in relations of difference, particularly economic, ethnic and gender differences.

In 2014, Thames Water in partnership with the environmental charity—Groundwork—initiated the Smarter Home Visit in two London Boroughs: Greenwich and Bexley. In advance of the introduction of the smart meter across the two boroughs, the aim was to provide households with expert advice on reducing water consumption and the offer of free water saving devices. Groundwork deployed 'Green Doctors' to carry out the household visits and install the devices. As the promotional material announced: 'As we roll out smart meters our team will be on hand to help you save water and energy—better for you, your pocket and your local river' (Thames Water website 2014). The intention was to help customers to get the most from their installed, or soon to be installed smart meter, with the promise of a potential saving of one quarter off the current water

3 Consuming Water: Habits, Rituals and State Interventions 57

bill. Once the smart meters have been installed and connected to the data hub, households can chart their appliance-by-appliance use of water on an hour-by-hour basis. This is a socio-technical assemblage of devices designed to shift household conduct which is being introduced in the context of a growing population and demand for water that is predicted to exceed existing sources of supply. During the visit, the smart meter is introduced to the often-bemused consumer as it is pulled out of the bucket along with the appliances; water saving devices are installed (Photo 3.1), advice on behaviour change to reduce consumption is given and water usage is estimated across a range of activities. What is being mobilized here is the figure of the average water consumer. But as Sofoulis (2011) points out, the average water user, and such statistical norms, 'smooth out complexity', while at the same time obscuring internal differences amongst householders and their practices, such that consumption is unequally distributed across generations, genders and ethnicities (Sofoulis and Williams 2008).

Photo 3.1 Green doctor anonymous. Sophie Watson

Other studies have also investigated similar programmes to consider behaviour change. For example, in their study of 252 households in South-East Queensland, Beal et al. (2013: 116) reported that 'householders' perceptions of their water use are often not well matched with their actual water use' and that 'attitudes and behaviour towards potable water supplies have changed due to greater social awareness and increasingly widespread exposure to drought conditions; people are beginning to genuinely value water as a precious resource'. My interest was in exploring the different ways in which such initiatives were mediated by sociocultural factors which I suggest affect the 'success' of the programme's stated objective to reduce household water consumption. My focus here also was on how the specific differences were constituted and articulated through the intervention. The households visited during the research period were spread across the two boroughs, with the largest proportion conducted in Thamesmead, an estate built initially by the Greater London Council for families who were rehoused from overcrowded Victorian terrace housing in inner London, and which now has a concentration of African households. Through thematic coding of the field notes taken during and after the visit, which recorded the willingness to have new devices installed, and the level of interest in the visit and the water saving advice given, two broad patterns of response were identified—engagement and resistance/indifference, where this latter response might be seen as an 'ambiguous and potentially productive process, rather than a deficient state of subjective failing' (Hynes 2016: 24). The responses were thus interrogated to explore the significance of cultural practices, rituals, gender/age/ethnicity, agencies, technologies and housing.

Affect

An analysis of discourses deployed by the Water Companies in the promotion of water saving programmes reveals explicit strategies to elicit an effective response. Customers are called upon to 'love every drop' (Anglian Water), 'love your river' (Thames Water), and 'care for the environment' and 'act responsibly', in attempts to mobilize environmental concern, awareness and an ethics of care (Gilligan 1982; Puig de la Bellacasa 2017).

3 Consuming Water: Habits, Rituals and State Interventions

What is often not recognized however, are the complexities inherent in any response to these messages, mediated as they are by social and cultural differences and histories. Most salient in this respect were income and education, gender, age, ethnicity and earlier experiences of Thames Water. None of these differences are homogenous, and differences themselves are produced through the smarter home visit rather than the existing pre-formed ones to be mobilized by the intervention. Ethnicity offers the clearest illustration of this. As we shall see, experiences of living with water scarcity or abundance, or unreliable or expensive water supply in other countries translates into different household practices in London.

Concern about the environment was more prevalent amongst households with higher levels of education, and amongst those in professional employment as other studies have found also (Gilg and Barr 2006). Many of these respondents already knew some aspects of water conservation and were keen to learn more and take the advice of the Green Doctors on water saving practices, and the use of devices. Respondents were typically expecting the 'smarter home visit', welcoming at the door, happy to answer the questions in the survey, asked questions themselves and were interested in how much water was saved by the installation of water saving devices. Concern was thus expressed in a willingness to change water use practices, a curiosity about water conservation and the reduction of water consumption. One householder for instance, said 'it is interesting to find out the various things you can do', reporting in the follow up interview that she found the visit both 'useful' and 'instructional' commenting: 'It did make me slightly more aware after the event of just general consumption and what you're doing during your day. But there's a minimum amount that goes on in the house that doesn't change' (Chinese businesswoman).

Several of the Green Doctors reported finding that some people with lower levels of education (as noted by Gilg and Barr 2006) were more reluctant to being educated about water use. A dominant trope here was a sense of marginality and powerlessness, which itself produces and reproduces new forms of marginalization where daily routines of survival become the limits of possibility. Green doctors mentioned the demoralizing effect of living in poor accommodation that worked against making changes, even if they enabled a reduction in expenditure. Lower income households, particularly those in private rented accommodation, were sometimes disengaged with the visit particularly where tenants were disempowered by their lack of rental security or permanence, or a marginal relationship to their housing which militated against being invested in the advice that was of little personal relevance. Where tenants were not responsible for the water bills, or the bills were paid on direct debit, there was a particular lack of concern about whether the devices could be fitted, or whether they helped save water. Equally in dwellings with a large number of occupants (often unrecorded), where secure tenure was precarious, there was a reluctance to discuss the bills or practices of the inhabitants.

Negative affect—expressed in indifference, resistance and lack of engagement with the initiative (60% of the recorded responses) was particularly notable amongst those who perceived Thames Water as an organization whose prime motivation was the accumulation of profits. These householders were suspicious of the motivations behind persuading customers to reduce their water consumption or change their practices:

> I don't think water should be for profit, obviously they have got to make some money to cover costs, but they shouldn't be making large profits for their shareholders. I tried to find out what profits they were making and they claimed crown immunity. (White middle-aged man)

And:

> If the devices did save water there would be benefits but then they push the prices up. I am concerned about the motives. I don't think it's a waste of time necessarily, but I am concerned about the motives. (Older white man)

Previous negative encounters with the corporation or local state for some householders also produced hostility or indifference to any external intervention. Herzfeld's (1992) book *The Social Production of Indifference* throws some light on this, where he mobilizes Weber's concept of 'secular theodicy' to refer to 'the idiom of grumbling against the state', which, he suggests, is deployed to justify earlier humiliation by the bureaucrats (1992: 127). In his view, although bureaucracy was intended to assure accountability, in reality it produces the opposite: indifference. A similar trend was identified by Browne et al. (2014) who suggested: 'if consumers regard their water provider to be untrustworthy, they are more likely to be unreceptive to proposed water conservation or efficiency initiatives, and thus these individuals (or households) are unlikely to be responsive to potential behaviour changes'. Suspicion underpinned other reactions where households assumed the Green Doctors were collecting information that could influence their bill. Where households were comprised of migrants who spoke little English, or felt threatened by strangers appearing at the door, household visits were declined or barely engaged with.

Habits

The smart meter initiative involves both education and the installation of new devices to reduce water consumption. Underlying such a strategy is the assumption that former habits and cultural practices enacted around domestic water can be changed. Such a notion, implicitly perhaps, involves an ABC framework of social change, where 'values and attitudes (the A) are believed to drive the kinds of behaviour (the B) that individuals choose (the C) to adopt (Shove 2010: 174), or an understanding of the 'nudgeable' human subject, who can be persuaded to change their behaviour (Thaler and Sustein 2009). It would be more useful, I suggest, to draw on the recent literatures on habit and on the interrelations between materials and objects, as they constitute our ways of doing things.

In their special issue of *Body and Society*, Bennett et al. (2013) argue for the centrality of habit in the formation of human capacities, and for understanding how particular behaviours and conduct are brought under the direction of both secular and religious authorities (2013: 4). As they put it: 'habit has more typically constituted a point of leverage for regulatory practices that seek to effect some realignment of the relations between different components of personhood—will, character, memory and instinct, for example—in order to bring about a specific end. Habit is always figured in relation to these other coordinates of personhood, caught up with them in processes of habituation, dis-habituation and re-habituation' (2013: 5). More recently, as they argue, there has been a move to understand the formation and re-formation of habits in their entanglement with socio-material environments. In Bourdieu, habit or rather—habitus as a set of dispositions—has been central to his accounts of how people relate to social worlds but takes little account of how material devices affect behaviour, which has led some Bourdieusian scholars to consider how physical environments and specific materials actively shape how fields are constituted (Silva 2016).

Strengers' (2013) exploration of smart technologies in everyday life and what happens when they encounter the household and the limitations and possibilities of smart strategies and materials in transforming and shifting everyday practices and routines, (following Shove et al. 2012) and reducing energy consumption, is very useful in opening up an understanding of everyday life as messy and disorderly (2013: 53). In relation to the willingness of people to adopt water saving devices, Marres (2012: 8–9) asks the question how 'things acquire the capacities to organize publics by particular means', which encouraged my attention during the smarter home visits to how the objects enrolled—or not—the householders in new cultural practices.

The devices were differently effective in challenging existing habits, which themselves were differentiated by ethnicity, age and gender. Mothers described the resistance of young women to the installation of the shower head to reduce time spent in the shower, since showering, and particularly the use of a plethora of body essences and creams, shampoos and conditioners, was seen as crucial to their daily routine to perform their bodies in ways that fit in with their peers. As Robbie (2009) argues, cultural expectations around beauty, body care and hair washing, proliferate through the media, which seems to translate into longer showers amongst teenage girls, some of whom reported showers of up to half an hour as common practice. As a consequence, one of the Green Doctors highlighted the benefits of a strategy which focused the smarter home visit on women with children: 'they are high water users—if we could hit them—then that is jack pot—that is my impression… girls do wash their hair every day—do their shaving—also enjoy the shower-… pampering … Definitely the teenagers—but also the 30 somethings are heavier users'.

Gendered and embodied habits translated into resistant responses to the intervention in diverse ways. Everyday water use for many of the lower income and single parent households was high, but the willingness to engage with new practices and use water saving devices was constrained

3 Consuming Water: Habits, Rituals and State Interventions 63

by the stresses and pressures associated with running a home single-handed. Single parents living in poor standard accommodation on benefits, on whom the pressures of bringing up children on often-limited incomes, not surprisingly, had little energy or time to engage with the visits. The lack of affordability of new washing machines or dishwashers with eco settings further militated against their future purchase, even if they understood the benefits of them. The use of washing machines was also similarly imbricated in cultural expectations, often gendered, around wearing clean clothes at work for example, and changing outfits daily. Women described taking pride in sending their children to school, or social and sporting events in freshly laundered outfits. As one woman put it: 'I'm not going to use my washing machine any less—my boys play football every day and I wash their clothes every time they come home.'

What is apparent from these interviews is not just that different people embody different 'cultural expectations' that override the desire to conserve, but that the very idea of what is a necessary versus a wasteful use of water is very culturally specific and varies across class and gender. In other words, no consensus exists around the notions of legitimate use, which themselves are cultural. The hydro-geography of the bioregion from which people originated, and experiences of hydro-politics (the various institutional and governance arrangements for water supply and demand management), whether rural or metropolitan, similarly seem to shape water habits, along an axis of distinct ethnicities and countries of origin, which constituted different cultural practices and habits. Migration from countries where there was a shortage of fresh water, or where water was used in very different ways, or managed differently, led to a strong interest in sharing their experience with water overseas and reflections on national and cultural differences. For instance, one man from the Democratic Republic of Congo commented that 'in England, people are more careless with water. In Kinshasa they already have water meters so people are more careful'.

A young woman from a mixed Afro-Caribbean and white British family described a background where water conservation was paramount. She made the point that her mother (a nurse) had been very insistent that they considered the environment when they were growing up. Keith, a computer analyst who was studying at Birkbeck College, and whose parents migrated from Jamaica, similarly reported being brought up not to be wasteful:

> I had my first trip to Jamaica when I was 21 and it was a bit of an eye opener—You had to pay for water there—to have the tank filled—so there you don't flush the toilet there unless you did no. 2—so it gave me a whole new look. We are a lot better off than most of my family out there. …I don't take it for granted any more.

In Thamesmead, where there is a concentration of African (particularly Nigerian) households, there was a pattern of using water sparingly, such as washing in the bath with a bucket or using mugs to brush teeth, practices even passed on to the next generation. A woman who had grown up in West Africa where water was scarce, after arriving in England, described carrying on the traditions from home, and was very enthusiastic about the devices, commenting on the shower timer: 'That will be good for my boys!'

As Tony, a Green Doctor, explained: 'Africans have had to walk to get water where they come from so they are not wasting it.' This kind of understanding was not only restricted to people from less developed parts of the world: 'My parents are in Australia and awareness of water consumption is very high there and so if you run a bath and there's cold water at the beginning you put a bucket underneath and save the cold water till it gets hot. You use it for something else—you don't just waste it. Behaviour is very much shaped by that awareness there and we don't have any of that in this country'. (White woman, 30s).

Religion also featured with sections of the Muslim community who articulated a cultural and religious based respect for water, and a tendency to conserve water, learnt from childhood. As Mohammed, a Bengali man explained: 'Islam taught me from an early age—when you go to Saudi-and you are on a pilgrimage and you use unnecessary water—you are told God won't be happy—even from childhood… Our religion says if you use more water for unnecessary reasons—it is seen as a sin….It is common sense'.

Yet, as we see in Chap. 7, the practice of Wudu (the ritual washing practice enacted before prayer by observant Muslims) disrupts this shared notion of what is appropriate or (un)necessary water use, since depending on where it is performed, a fair amount of water might be used.

The installation of swivel taps in the kitchen sink, or the suggested use of a washing up bowl in the sink, were resisted by several ethnic groups. One Nepalese man described the difficulty of persuading the women in his household to wash their vegetables and dishes in a bowl, since in his country fresh running water was abundant in the many rivers that descended from the Himalayas, and the notion of washing things in water that was not flowing was thus seen as dirty and unhealthy. Some of the Asian women interviewed considered washing vegetables or dishes in static water to be unhygienic. As one woman in a large suburban house explained to me: 'the dirt just gets recycled and absorbed into the vegetables—it's a horrid idea'. Others came from countries where water was free, such as in Turkey, which meant that they had little awareness of its cost, even thinking it was free, with the result that they were not mindful of taking care to avoid high water bills.

It is clear from these responses individuals were not easily enrolled into the initiative for a diverse array of reasons, produced within a complex assemblage of factors (as Jenkins and Pericli 2014: 58 also found) which were not reducible to lack of motivation or other negative accounts of individual behaviours.

Some older people who were more careful with resources and better at conserving energy and water, as other studies found, typically embraced the devices. The generation who grew up during the Second World War and the post-war period remains imbued with a sense of responsibility for scarce resources and an abhorrence of waste, particularly women, whose primary domestic responsibility brought heightened awareness. As Miles, a Green Doctor, pointed out:

> The older people particularly are very conscious of these things. Taking water saving advice to people who are 50 plus they can tell you what to do. The younger generation much less so. Older people who grew up in the war pass on advice to the baby boomers—it's the next generation who are spoilt with everything.

But this was by no means universal. Age intersected with established habits perceived as not amenable to change. Resistance to instructions determining their personal habits echoed the responses of the participants in the Everyday Water research in Sydney (Allon and Sofoulis 2006:

53), who were 'highly critical of the kinds of scripts that some technologies in the home dictate for users to follow'. As one older white couple put it: 'we are too old to change our behaviour now and why should we be told to do so', or as Bob, one of the Green Doctors said, 'it was difficult to teach old dogs new tricks—an older person who has always had baths—not going to change them—I like my bath—a younger person might be up for a two minute showers'.

Faulty devices were a matter of concern. In the follow up interviews where devices had been installed, if they had functioned well, householders were more likely to have been successfully enrolled in the initiative. But rising costs of water also mobilized antagonism and annoyance as one older white man explained:

> I used to keep koi carp—which I gave up 6 years ago when I had a serious accident and also when the neighbour poisoned my fish—and at that time I used 6 cubic meters of water. But now I get even higher bills even though I am using much less water obviously. 6 years ago my bill was £85 per half year, now it is £186 per half year. So clearly they have put the prices up.

And another middle-aged white man:

> I think it is all about PR. Water should not be provided by profit-based organisations. Also I am a technical person and this aerator they put on the kitchen tap—it is failing already. It is leaking—letting water through. And it has only been on for two months. It is also getting furred up as we have soft water here. So I am thinking of taking it off.... They put in the saving device in the cistern. I think they could more usefully do a simple conversion to dual flush—it would cost £20—and that would work far better. You then use 3 or 1 gallon.

Domestic architectures and housing tenures of households intersected with devices. Where houses were large and luxurious, saving money was not a great concern, or wealthy owners were not interested in unsettling their carefully thought through domestic design with objects not of their choice. These householders acknowledged their high water bills, often deriving from several bathrooms and facilities (one house had seven bathrooms, nine taps and two dishwashers), but had little motivation to reduce their bills. As one Green doctor Bill described it:

I went to a man who didn't care that much. He had a nice home—I offered a shower timer—he said "no don't bother—if I want a 10-minute shower I'll have one"—so saving £50 on shower is meaningless to them.

Meaning of the Home

Finally, this study of the intervention revealed that water authorities fail to take account of the specificity of meanings of home, in particular the notion of home as private and free from regulation and intervention—a space of personal control. Thus, what is not recognized is that the smarter home initiative disrupts and reconfigures the meaning of the home in particular ways. The point here is that domestic water consumption is integrally connected to particular practices deemed appropriate to the 'home' or to assumptions, beliefs and values about cleanliness, comfort and convenience—Shove's 3 Cs' (2003: 3) which encompass the environmental hot spots of consumption'. These everyday practices (washing, going to the toilet, showering), which are embodied, saturated in affect, habitual and entrenched over time, are difficult to shift. They play an important role also in maintaining social and familial relations and are implicated in notions of the home that are contested, shifting and contextual (Allon and Sofoulis 2006). Nevertheless, in advanced capitalist societies homes are typically seen as private, as spaces of belonging, the familial, and personal control, though these attributes are highly mediated by tenure and gender also (Watson 2010).

Feminist writers have long articulated private/public boundaries as shifting, contextual and complex, where activities associated with the public sphere are enacted privately, and vice versa. That said, private-public boundaries retain some force, both imaginary and symbolic, as a respite from the troubled and hectic world outside, and material—the door can be locked to keep outsiders at bay—as is starkly evident in gated communities. Thus, the smarter home visit, in its material practices enacted, reconfigures the public/private boundary by making the home a space of public intervention and by bringing normally private embodied matters into public discourse and regulation: tooth brushing, time spent in the shower and practices of defecation. This is not normally the talk or terrain of strangers. The research confirmed this point. In the household visits, we witnessed awkwardness and embarrassment at questions relating to personal habits.

Similarly, the fitting of the smart devices disrupts notions of the home as a site of individual choices over objects like taps, showers and washing machines. In a culture where homes and the objects within them are key expressions of self, 'a space of belonging and alienation, intimacy and violence, desire and fear, … invested with meanings, emotions, experiences and relationships that lie at the heart of human life' (Blunt and Varley 2004: 3), such interventions are similarly imbued affectively. In the research, several respondents said they had chosen their fitting to suit their taste and could not be persuaded to adopt alternatives even if they saved water. The smarter home intervention thus disrupts and reconfigures private-public boundaries exposing private behaviours to public scrutiny and intervening to reconfigure domestic bodies as responsible subjects who are required to shift habitual and desired habits to conform to wider public objectives. The devices are also actors which reconfigure the private as a space where water users are enrolled in new ways of relating to water, over which they have limited control. For example, the saver flush inserted into the cistern reduces the amount of water in the toilet, which militates against large organic or inorganic objects being washed away. The new showerheads reduce water flow from the more typical 15 and 20 litres flow per minute to 8 litres per minute, diminishing the pleasures of revelling in an abundant shower. Though environmentally practical, these socio-technical assemblages are thus constitutive of new embodied practices re-figuring homes as sites of intervention.

What is ignored therefore in smarter water initiatives is that the particular meanings of home, which vary across cultures, time and place, act to configure particular responses of households and the individuals within them, and social differences. What these interventions do is take control away from the households in ways that are challenging and liable to be resisted. Such resistance was revealed in some of the follow up interviews. One of the women described throwing away many of the devices since she considered it her business to run the household in the way she wanted to and resented the intervention of outsiders telling her how to run her home. This was a particularly gendered response, most evident when women took the major responsibility for domestic life.

Conclusion

Everyone needs water survive. Without water humans, animals and crops will die. Yet water is a resource that is used profligately in many cities of the world, while in other cities people have no fresh water to drink and are reliant on water delivered in bottles, or on one tap shared with many. Water is the cause of great inequalities across the globe. How much water societies actually need is open to contestation and shifts across time and space according to changing cultural, social, industrial and agricultural practices. What is clear, however, is that with climate change and growing populations in many parts of the world water is becoming a scarcer resource with the need for better management and practices of delivery and changing practices of consumption. The fact that more water is needed to produce meat for humans to eat, than to produce many vegetable crops, for example, means that our future reliance on meat as a major source of nutrition may well have to be shifted.

We saw in this chapter how water consumption is embedded in a complex assemblage of socio-technical connections, which are not necessarily given but which are constituted in different ways at different times. It is also produced within a complex array of unremarkable habits, practices, rituals and conventions which differ from place to place, and which are imbued with specific cultural meanings. These too change over time and across space, with daily routines shifting as a result. Similarly, new technologies engender new everyday practices and politics which themselves disrupt the boundaries of private and public life. In this sense, understanding water consumption practices requires exploring the changing materialities, cultures, politics and infrastructures of public life as this chapter has demonstrated.

References

Allon, Fiona, and Zoe Sofoulis. 2006. Everyday Water: Cultures in Transition. *Australian Geographer* 37 (1): 45–55.

Anand, Nikhil. 2015. Leaky States: Water Audits, Ignorance, and the Politics of Infrastructure. *Public Culture* 27 (2): 305–330.

Anderson, Ben. 2009. Affective Atmospheres. *Emotion, Space and Society* 2 (2): 77–81.

Beal, Cara, Rodney Stewart, and Kelly Fielding. 2013. A Novel Mixed Method Smart Metering Approach to Reconciling Differences between Perceived and Actual Residential End Use. *Journal of Cleaner Production* 60: 116–128.

Bennett, Tony, Francis Dodsworth, Greg Noble, Mary Poovey, and Megan Watkins. 2013. Habit and Habituation: Governance and the Social. *Body and Society* 19 (2&3): 3–29.

Blunt, Alison, and Ann Varley. 2004. Geographies of Home. *Cultural Geographies* 11 (1): 3–6.

Browne, Alison, Will Medd, Martin Pullinger, and Ben Anderson. 2014. Distributed Demand and the Sociology of Water Efficiency. In *Water Efficiency in Buildings: Theory and Practice*, ed. K. Adeyeye. Chichester: Wiley Blackwell.

Carrington, Damian. 2017. Thames Water Given Maximum £8.5m Fine for Missing Leak Target. *The Guardian*, June 14.

Gilg, Andrew, and Stuart Barr. 2006. Behavioural Attitudes Towards Water Saving? Evidence from a Study of Environmental Actions. *Ecological Economics* 57: 400–414.

Gilligan, Carol. 1982. *In a Different Voice: Psychological Theory and Women's Development*. Boston: Harvard University Press.

Hawkins, Gay. 2011a. Commentary. *Environment and Planning A* 43: 2001–2006.

———. 2011b. Packaging Water: Plastic Bottles as Market and Public Devices. *Economy and Society* 40 (4): 534–552.

Hawkins, Gay, Emily Potter, and Kane Race. 2015. *Plastic Water: The Social and Material Life of Bottled Water*. Cambridge, MA: MIT Press.

Head, Lesley. 2008. Nature, Networks and Desire: Changing Cultures of Water in Australia. In *Troubled Waters: Confronting the Water Crisis in Australia's Cities*, ed. P. Troy, 67–80. Canberra: Australian National University Press.

Head, Lesley, and Pat Muir. 2007. Changing Cultures of Eastern Australian Backyard Gardens. *Social and Cultural Geography* 8 (6): 889–905.

Herzfeld, Michael. 1992. *The Social Production of Indifference*. Chicago: University of Chicago Press.

Hynes, Maria. 2016. Indifferent by Nature: A Post-humanist Reframing of the Problem of Indifference. *Environment and Planning A* 48 (2): 24–39.

Jenkins, James, and Alexis Pericli. 2014. Understanding Consumer Response to Water Efficiency Strategies. In *Water Efficiency in Buildings: Theory and Practice*, ed. K. Adeyeye, 61–73. Chichester: Wiley Blackwell.

Maller, C., and Yolande Strengers. 2013. The Global Migration of Everyday Life: Investigating the Practice Memories of Australian Migrants. *Geoforum* 44: 243–255.

Marres, Noortje. 2012. *Material Participation: Technology, the Environment and Everyday Publics*. London: Palgrave Macmillan.

McFarlane, Colin. 2013. Metabolic Inequalities in Mumbai. *City* 17 (4): 498–503.

Puig de la Bellacasa, Maria. 2017. *Matters of Care. Speculative Ethics in More Than Human Worlds*. Minnesota: University of Minnesota Press.

Randolph, Bill, and Patrick Troy. 2008. Attitudes to Conservation and Water Consumption. *Environmental Science and Policy* 11: 441–455.

Robbie, Angela. 2009. *The Aftermath of Feminism: Gender, Culture, and Social Change*. London: Sage.

Robinson, D., K. Adeyeye, D. Madgwick, and A. Church. 2014. *Beyond the Water Efficiency Calculators*. Edited by B. Smyth et al. Brighton: University of Brighton.

Ruvaga, Lenny. 2015. Business Farming Go Greener to Protect Nairobi Water Supply, March 23. https://www.voanews.com/a/kenya-turkana-water-unfit-for-human-consumption/2674350.html.

Sahoutara, Naheem. 2017. Karachi's Water Unfit for Human Consumption. *The Express Tribune*, July 15, Sindh, Pakistan.

Shove, E. 2003. *Comfort, Cleanliness and Convenience: The Social Organization of Normality*. Oxford: Berg.

Shove, Elizabeth. 2010. Beyond the ABC: Climate Change Policy and Theories of Social Change. *Environment and Planning A* 42: 1273–1285.

Shove, Elizabeth, Mika Pantzar, and Matt Watson. 2012. *The Dynamics of Social Practice: Everyday Life and How It Changes*. London: Sage.

Silva, Elizabeth. 2016. Unity and Fragmentation of the Habitus. *The Sociological Review* 64 (1): 166–183.

Sofoulis, Zoe. 2005. Big Water, Everyday Water: A Sociotechnical Perspective. *Continuum: Journal of Media and Cultural Studies* 9 (4): 445–463.

———. 2011. Skirting Complexity: The Retarding Quest for the Average Water User. *Continuum: Journal of Media and Cultural Studies* 25 (6): 795–810.

———. 2013. Below the Double Bottom Line: The Challenge of Socially Sustainable Urban Water Strategies. *Australian Journal of Water Resources* 17 (2): 211–221.

Sofoulis, Zoë, and Carolyn Williams. 2008. From Pushing Atoms to Growing Networks: Cultural Innovation and Co-evolution in Urban Water Conservation. *Social Alternatives* 27 (3): 50–57.

Strengers, Yolande. 2013. *Smart Energy Technologies in Everyday Life: Smart Utopia*. Basingstoke: Palgrave Macmillan.
Taylor, Vanessa, and Frank Trentmann. 2011. Liquid Politics: Water and the Politics of Everyday Life in the Modern City. *Past Present* 213 (1): 199–241.
Thaler, Richard, and Cass Sunstein. 2009. *Nudge: Improving Decisions about Health, Wealth and Happiness*. London: Penguin.
Thames Water. 2014. http://www.thameswater.co.uk/metering/17037.htm.
Thomson, John. 1877/1994. *Victorian London Street Life*. New York: Dover Publications. Originally published 1877. London: S. Low Marston, Searle and Rivington.
Vanolo, Alberto. 2014. Smartmentality: The Smart City as Disciplinary Strategy. *Urban Studies* 51 (5): 883–898.
Walker, Anna. 2009. *Independent Review of Charging for Household Water and Sewerage Services*. London: DEFRA.
Watson, Sophie. 2010. City A/Genders. In *The New Blackwell City Reader*, ed. Gary Bridge and Sophie Watson, 237–243. Oxford: Blackwell.
WHO/UNICEF. 2018. *WHO/UNICEF Joint Monitoring Programme (JMP) for Water Supply, Sanitation and Hygiene*. New York: United Nations.

4

River Powers: Assembling Publics, Connections and Materials in a Global City

A cursory glance at any atlas reveals the significance of rivers to the construction of many of the world's capital, and subsidiary, cities, many of which are also close to the sea. Most capital cities are situated on rivers or harbours, and are central to the city's identity and image. Rivers enable commerce, trading, industry, irrigation, food production and water to drink, and for many centuries have represented the lifeblood of the city, quite literally its major artery. Taking London's Thames as its prime illustration, this chapter considers the co-constitution of rivers and the cities through which they flow. It seeks to illustrate the liveliness of rivers in making urban everyday life and textures, where their very materiality and form—the tides, the mud and the water—all have effects which produce different forms of work, built structures, circulation, and pleasures which have shifted over time. River technologies, boats, craft, docks and wharves similarly produce, and are produced by, shifts in everyday cultural practices and technical—more recently digital—knowledges and economic imperatives. In these ways rivers enrol cities and their populations in a complexity of changing relations and with changing meanings. Reading a city through the river that runs through it reveals new understandings and tells different stories. As Coates (2013: 15) suggests, rivers are 'the collective product, not just of geology, ecology and climate, but of eco-

© The Author(s) 2019
S. Watson, *City Water Matters*, https://doi.org/10.1007/978-981-13-7892-8_4

nomics, technology, politics and human imaginings…London has been dubbed the gift of the Thames'.

The importance of rivers to city cultures, politics and economies has ebbed and flowed. In 1861 Henry Mayhew, in his *The Illustrated Mayhew's London* began his account of the river

> by enumerating the numerous classes of labourers, amounting to many thousands, who get their living by plying their respective avocations on the river, and who constitute the customers of these men. These are first the sailors on board the corn, coal, and timber ships; the 'stevedores', or those engaged in stowing craft; and the 'riggers', or those engaged in rigging them; the ballast-heavers, ballast—getters, corn-porters, coal-whippers, watermen and lightermen, and coal-porters.

This description evokes the crowded and animated life of the Thames, represented in books, paintings, and photographs of London from its earliest days until the post-war period of decline. This was a material culture of boats, and objects where 'steamers pass and repass, twisting and wriggling their way through craft of every description, the unskilful adventurer would run into continual danger of having his boat crushed like a shell'. The river was so full of boats that at certain moments the foolhardy youth could cross from one side to the other hopping from deck to deck without falling in. The artist J.M.W. Turner (1775–1851) was born near the river, and unlike other painters of the time his paintings depict the smoke and fog of the river; and, reflecting his fascination with modernity and industrialization, his painting of *The Thames Above Waterloo Bridge* shows the city's factories and river traffic producing fumes which virtually obscure the bridge.[1] In an interview, Martin Garside at the Port of London Authority (PLA) recounted an amusing story about Turner's love of the river:

> Turner is interesting because he painted both ends of the river because he lived up at the west part and then he did these amazing sunsets at the estuary at Margate. You know the amusing story about that, don't you?… The original art histories would say, 'This is because he was drawn to Margate by the amazing sunsets and the fact the sun sets over the estuary', which indeed it does. But do you know there was another reason why he went

---

[1] http://www.tate.org.uk/whats-on/tate-britain/exhibition/turner-whistler-monet/room-guide/room-5-turner-and-thames.

there?…some landlady he was rather fond of at Margate. But he also painted up river as well. That's Turner for you.

The Thames was not only depicted in paintings, it was equally the stage on which much literature was set, most notably in Dickens's (1865) *Our Mutual Friend* which opens with the description of a 'boat of dirty and disreputable appearance, with two figures in it, floated on the Thames between Southwark bridge which is of iron and London Bridge which is of stone, as an autumn evening was closing'. Here the river is a threatening and dark place, beset with deaths and drownings, where, for example, Lizzie and Gaffer Hexam make their living by trawling the river for victims and emptying their pockets before handing the bodies over to the police.

Mid-Nineteenth Century to Mid-Twentieth Century

It is only possible to briefly depict several moments in the long history of London's vibrant river. The Thames witnessed a dramatic increase in traffic and use as the British Empire extended its reach, through international trade and expanding markets, larger ships, and an increasing population in the capital city. In 1802 the West India Dock was opened by the Prime Minister as a result of a long-standing campaign by the merchants trading in the West Indies, to provide relief from the growing congestion in the port. The key figure in its construction was the merchant Robert Milligan, who had returned from managing his family's sugar plantations in Jamaica. Like many of these early traders he was a slave owner who made his fortunes exploiting the labour of 526 enslaved people—though this is a story often obscured in celebrated histories of the merchants of this time.[2] This was followed shortly by the construction of the London Dock (1805), the East India Dock (1806)—predominantly trading in tea, spices, cotton and silk—and St Katharine Dock (1828). This was a material culture where buildings represented London's imperial power through the grandeur of the built structures straddling the river across its tidal reach. Wharves grew up along the river

---

[2] https://www.ucl.ac.uk/lbs/person/view/2146645741.

to handle goods. As the century progressed more docks were built—Victoria Dock (1855), Millwall Dock (1868), South West India Dock (1870), Royal Albert Dock (1880) and Tilbury Dock (1886) to manage the new and larger steamships trading in grain, timber and foods from the colonies (Steadman Jones 1971). These few decades were a notable example of the free market investment and infrastructure undertaken by Victorian entrepreneurs, which marked the river as a space of commerce and trade constituted by, and constitutive of, unequal and racialized power relations at home and abroad.

'Workers on the Silent Highway' (Thomson 1877)

For centuries the Thames was the focal point of London life where many of the men in the capital found employment: Stow's (1876) survey of London in 1598 related that some 40,000 men earned a living on or about the river. The variety of jobs to be found there is evoked in Henry Mayhew's description mentioned earlier. Thousands of stevedores—a term derived from the Portuguese 'estivador'—and more recently known as dockers, were on hand daily to load and unload the numerous ships departing to, and arriving from, distant parts of the world. Watermen were in charge of the wherries which took people from one place on the river to another, and the lightermen were responsible for moving the freight on the barges. Henry Gosling (1927: 23–9), who started life on the river following a long ancestry of watermen and who became Ramsay Macdonald's first Labour Minister of Transport, described life on the river as never monotonous, but as requiring 'men who accept fog and frost and snow and sunshine…as part of a day's work'.

The watermen's work was constantly shifting and changing, as the built environment, innovations in transport and finally the demise of the docks in the 1970s–1980s threatened their livelihoods. As early as the beginning of the seventeenth century, watermen were defending their trade from the threat brought about by the construction of theatres away from the South side of the river in the West End, which was closer to the concentration of residential population, which was followed by a resistance to the introduction of the hackney carriage in 1634 which was undermining their monopoly. James from the Port of London Authority commented: 'because they were the cab drivers of the seventeenth century and one suspects there's a certain common ground with the cab drivers who are now raging against the dying of the light with Uber. Who were the people who've campaigned most against bridges? It was the waterman, because it would kill the trade.' Two centuries later, Thomson (1877) describes the threat to the watermen's employment and prestige, as the 'silent highway' ceased to be the main thoroughfare of London life and commerce, superseded as it was by the construction of smooth streets and by cab, bus, steamer and train transport.

But these were, and remain in some respects, protected and esteemed occupations, which in part can be attributed to the power and prestige of the Company of Watermen and Lightermen. For centuries the river had existed as an unregulated space. As Iain Reid of the Company of Watermen and Lightermen described:

> Prior to the Act it had all been small ferryboats and basically guys in a classic 'man for all seasons' style, you hail a wherry in Hampton Court and things like that, but they were particularly renowned for being rather rough and ready and not really always abiding by the rules and you'd be half way across the river and then they'd start to renegotiate the price. And there was quite a lot of underhand … there are a number of stories… and one of the popular stories around the City is that at the time of the Great Fire of London the watermen doubled their rates to evacuate people from the City.

In 1514 a Parliamentary Act was thus introduced to regulate the fares charged on the Thames, which was initially ignored by the watermen carrying passengers independently (Photo 4.1). In 1855 a further Act of 1555 appointed 'Rulers of All Watermen and Wherrymen' working between Gravesend and Windsor, leading to the formation of the Company. The Act also introduced apprenticeships for a term of one year for all boys who wanted to learn the watermen's trade. The lightermen—those carrying goods and cargo—joined the Company in 1700 and in 1827 the Company was incorporated by a further Act of Parliament which ensured its independence. Since then it has been governed by a Court of Assistants, which includes an annually elected Master together with four

Photo 4.1 A group of Watermen

Wardens. The Company continues to enact a symbolic and material power over the Thames in a Georgian building designed by William Blackburn in the City dating from 1780. It remains as a working guild, providing services to its Freemen, facilitating an apprenticeship scheme, acting as trustee for its charities and participating in the various traditions of the City of London. The Company has over 390 Craft-Owning Freemen and also some 500 Journeymen Freemen who have completed a five-year apprenticeship to become qualified Watermen and/or Lightermen (now the Boat master's Licence). This training involves considerable skill and memory, since the Thames as a tidal and bendy river, crossed by many bridges of varying heights, is far from easy to navigate safely. Iain Reid described the final examination still undertaken this way:

> It's a bit like cab drivers doing the knowledge. So the apprentices come in, dressed up in their best suit for the first time they've worn it, and I sit in

4 River Powers: Assembling Publics, Connections and Materials…

occasionally although I don't contribute very much… there's a long table, it's a bit like doing a viva. You come in but you stand in front of a long table upstairs with about seven or eight people sitting behind it who will then say, 'Right, you're setting off from Gravesend pier at night. It's a flood tide. Go through all the navigation lights' and 'You're going to Putney, can you go through all the lights and signals you will see?' And then particular ones they'll say what's the particular note on a particular reach, and you've got to know that there are outflow pipes from the power station or that there's a disused jetty which is part submerged by day.

Dean Hill, a lighterman on the Thames today, from a lighterman's family stretching back over generations, described the 'scary' day he went for his license (Photo 4.2):

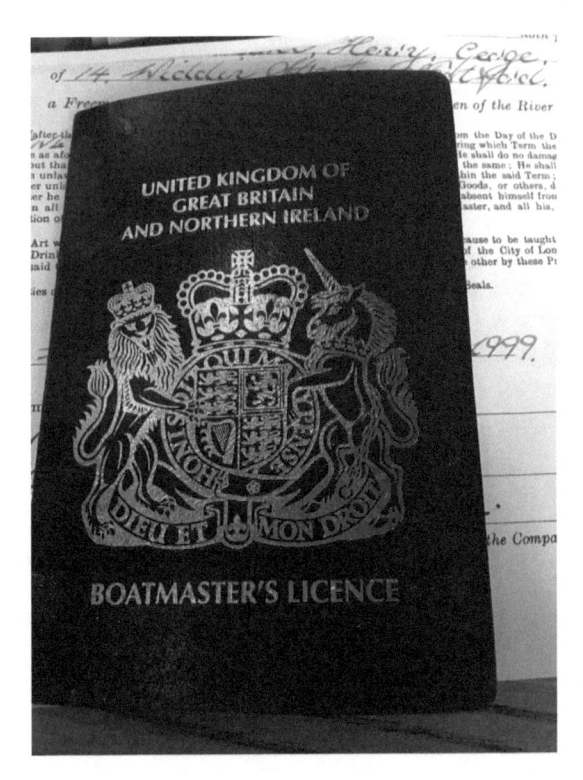

Photo 4.2 A waterman's licence. Sophie Watson

I thought I was going to a job interview, it's like a whole, big ceremony thing and you have to swear allegiance to the Queen, they make you sign these forms see how it's split in half there—one half belongs to your master, which was my grandfather, and the other half is mine. This is from like… the Watermen's Company came about in about fifteen hundred and something.

The powerful aura of the Company and its demanding apprenticeship was an important part of life on the river through the centuries as Gosling in 1927 (16) explained:

I do not think any Thames waterman is ever likely to forget the day on which he was bound apprentice. …everything conspired to give dignity and solemnity to the binding act. …The expected codes of behavior were strict, if unlikely to be adhered to. Every apprentice pledged himself to … 'not commit fornication, nor contract matrimony within said term.: he shall not play at cards, dice, tables.'

Like many city guilds, the Company remains steeped in traditions. One of these is the award of Doggett's coat and badge to the best apprentice which was first awarded 300 years ago. Iain Reid explained its origin:

There's hundreds of stories as to why he did this. Thomas Doggett (1640–1721) was an actor and the rumour is that one night trying to get back from Drury Lane or somewhere it was pouring with rain and everybody refused to take him except somebody who was just coming out of his apprenticeship and so he awarded what is known as the Race for Doggett's Coat and Badge, there's been a winner every year for the last 300 years.

Mudlarks, Sewer Hunters, Dredgers and Purl Men

Aside from the more established, albeit by no means always secure, employment, there were many other people deriving their income from the river in more tangential ways in often the most liminal of sites. These

4 River Powers: Assembling Publics, Connections and Materials…

included the mudlarks, sewer hunters and dredgers, who trawled the river for discarded objects which could be sold on for a profit. This was by no means a homogenous group of marginal labourers, rather, as Mayhew (1861: 189) points out, there were fine distinctions both in the money earned and in their everyday work and cultural practices, with surprising hierarchies established across the groups. This was also an arena of work where women were involved:

> The sewer-hunters are 'a far more intelligent and adventurous class (than the bone—grubbers or mud-larks), but they work in gangs. They must be familiar with the course of the tides, or they might be drowned at high water' …The dredgermen, the finders of the water, are again distinct, being watermen and working in boats… they also depend for their maintenance upon the sale of what they find, or the rewards they receive.'

As Mayhew (1861: 199) underlines, it is surprising that men could be found who were prepared 'for the chance of obtaining a living of some sort or other, would, day after day, and year after year, continue to travel through these underground channels for offscouring of the city'. Yet far from receiving a negligible income these men made 6 shillings a day in their discoveries of coins and jewellery, which was more than many artisans at the time, but which according to Mayhew they drank away in the public houses at night (who would blame them!).

The mudlarks were at the bottom of the pile (Mayhew 1861: 206–14), scraping in the mud at the side of the river for whatever they could find—and as with so much degraded and marginal employment, this was typically women's (and children's) work, where earnings were minimal (two and half to eight pennies a day) and which involved the collection of old iron, bones, rope and copper nails (the most valuable).

This was a tough and miserable daily life where in order 'to obtain the articles they seek, to wade sometimes up to their middle through the mud left on the shore by the retiring tide….they may be seen of all ages, from mere childhood to positive decrepitude, crawling along the barges at the various wharfs along the river…their bodies are grimed with the foul soil of the river… among (them) may be seen old women…bent nearly double with age and infirmity'.

In a scathing dismissal Mayhew (1861: 207) describes them as 'dull, and apparently stupid'.

To describe life on the river as devoid of pleasures would be to dismiss the indulgences that any generation of workers seek to ease their troubled life and mitigate the degradations of often alienated and arduous labour. Beer drinking was one diversion, which took place in public houses, but when time was short, and work was arduous, was provided on the river by the Purl Men (ibid.: 158–68), who brought the beer to the river men as they worked. Purl was a mixture of beer infused with wormwood which was warmed up to boiling and infused with gin, sugar and ginger and taken by skiff to the boats where the purl man announced his arrival with a noisy bell. To establish yourself as a purl man it was necessary to procure a license from Waterman's Hall at 3s 6d per annum, since to navigate the waters through the river crowded with steamers and other craft involved considerable skill and practice.

Unruly Waters

> Kingdoms may come, kingdoms may go, whatever the end may be;
> Old Father Thames keeps rolling on, down to the mighty sea.
> Gracie Fields (1930s)

> Old father Thames advanced his reverend head;
> his tresses dropp'd with dews, and o'er the stream
> his shining horns diffused a golden gleam:
> grav'd on his arm appear'd the moon that guides
> his swelling waters and alternate tides: Alexander Pope (1713).

The Thames is a tidal river rising and falling twice a day between five and seven metres. It is also a river of meandering loops and bends, where the tides create complex currents and flows of water, of which watermen are apprised in their training. Such shifts in large volumes of water have huge effects on the riverbanks and structures, as well as on the navigation of craft. This is an unruly river subject to unexpected movements of water, as it makes its way from the sea to Teddington, a name derived from the Saxon Tides End Town, to the sea.

4 River Powers: Assembling Publics, Connections and Materials...

Martin at the PLA described the river's power this way:

> I mean you only have to watch East Enders, don't you, each evening to see how bendy it is. You've got aggressive currents, five or six, four or five knots is the tidal flow from the sea and that's in and out, you've got lots of structures which man in his wisdom has decided to build, each bridge is a challenge. If you think about it each bridge has been built almost randomly, that isn't quite the right word, but they haven't been built with the seafarer in mind, have they, they've been built because they look magnificent or to serve a functional purpose or to get from A to B and they've been built, well, the oldest bridges in London are well over one hundred years old. Can you imagine it? You do not have clear line of sight.

For centuries the movement of boats and vessels proceeded uncontrolled, leaving the intricacies of its navigation to the skill of the men on the river. In 1857 the Thames Conservancy was established to control and regulate the river. What was needed were initiatives to improve navigation in the river through ensuring an adequate depth for vessels to move freely, to rebuild locks and weirs and introduce other such measures to take care of a long abandoned waterway. Richmond lock was one such impressive Victorian engineering design, constructed to respond to two sources of water, the huge rush of water from the sea and the fluvial flows coming down from its source, and working perfectly more than 100 years later. This was a period of rapid technological change, when all the locks on the river were converted to electric or hydraulic navigation. In 1908, war in Europe was on the horizon, and London's place at the centre of a large and far-reaching empire had extended its multiple trade links across the world. Everyday working life at the docks was far from calm as the docks engaged in fierce and cutthroat competition with one another, thefts and pilfering were rife, and labour unrest was common. Under the instigation of Lloyd George and Winston Churchill, Parliament saw the necessity of establishing a single authority to regulate and manage the tidal river from the estuary to Teddington. The Port of London Act was passed in 1908 delegating the ownership of the land, the management of cargo and the employment of the majority of dock labour to the new body. A PLA police force was established to challenge the then high levels of pilferage—depicted amusingly in a Museum of London Docklands photograph of police officers jumping into the river to test their cork life jackets.

As a public trust it was accountable to Parliament and received no funding from tax revenue, relying on income from charges, where any surplus was retained and reinvested into port facilities. Its power is strikingly materialized in their magnificent and imposing building in Trinity Square. All the income was derived from activity on the river, in the form of small levies and charges on all commercial operations, such as on cargo, on each ship arrival and so on, and licenses for works in the river, a model which continues to the present day. Over the course of the twentieth century the PLA was also handed management of the Victorian docks, which they extended, representing a source of further income, until their rapid demise with dock closure during the 1980s.

The Port of London Authority thrived from the early to mid-twentieth century, playing an active role during the two wars until the 1970s when a revolution in world shipping logistics and huge structural change in the shipping sector through the advance of containerization—The *C word*, as Martin put it, brought dramatic changes to the docks. For ship owners the imperative was to facilitate the movement of cargo up and down the English Channel and the North Sea, and to be closer to the main shipping routes. The construction of large containers necessitating bigger ships and terminals to receive them, made the docks up the river obsolete. Extensive warehouses in the centre of cities were no longer needed. Other trends in logistics compounded the impact, such as the development of orbital motorways like the M25, which further detracted from the need for central London warehousing. The expansion of roll-on roll-off ferries enabled a more efficient method of moving lorries that arrived from the continent and became core to the activity of the modernizing Thames. A Belgium-based company Cobelfret, for example, operated six or seven services a day by 2017, carrying huge amounts of cargo, none of which could go further up the Thames than Purfleet.

Martin explained:

> This is very simplistic and also broadly accurate, so the Victorians brought their ships into the centre of the capital, they built large warehouse facilities,

4 River Powers: Assembling Publics, Connections and Materials…

ships more or less unloaded by hand, and that was going on… OK, they used winches and trollies and things, but it was very labour intensive. You then loaded it to warehouses and then they were put on small lorries and delivered round the London area. That was the model which worked very successfully, I mean as late as the early 1960s London was absolutely buzzing and thriving. Then containerisation arrived.

Massive unemployment, redundancies and labour unrest followed. The docklands transformed from a vibrant space of employment, trade and a buzzing everyday life, where noise filled the air and workers spilled out from the many public houses on the waterfront at night, to a series of wastelands in the space of just a few years. Dean Hill, a lighterman on the Thames, described those days:

> It was great, at that time everyone who I worked with were young lads. I was, like I say, 18, the captain of the boat was about 23, we was all youngsters, so there was a real community feel. And then when you used to go in the docks where my granddad used to work in the dock with his mates, like the older lot would be in the dock, loading and unloading the barges, with their little tiny tugs that used to bring all the little boats around, we was on the big one, used to take it down to the Medway. And then every now and again you'd get back to West India Dock and you'd have four hours where you didn't have to go again, so you would all go in the pub called The Gun.

In the space of just a few years, only one large Victorian dock remained—Tilbury, which had limited commercial potential when it opened in the late Victorian era, but whose construction was prescient in the long term, as it was the only dock able to offer a large container birth which the containership companies need.

Martin again:

> Well, the closure of the Docks happened very, very quickly. …it was very, very sudden. And then Tilbury Dock became this booming dock but mainly with the containers. And so there wasn't the foresight, unfortunately, to secure some of the old riverside wharves up river where there would've been some great connectivity, because clearly containers can be carried by a barge, we could've had bigger barges lightering the bigger ships

down in the Estuary and bringing them up no different to how Cory Environmental do today with rubbish. So it could've happened. Unfortunately, there wasn't that foresight, the road was so much more popular… it costs more money now to transport a container from Felixstowe to the Midlands than it does to get it from China to Felixstowe.

In 1981 the Thatcher Government set up the London Docklands Development Corporation to regenerate the Docklands area and encourage huge investment amidst intense political controversy. Employees of the PLA fell from tens of thousands to its present workforce of 350 staff. The carefully constructed wharves, which were owned by modest family companies and private entrepreneurs often dating from many centuries earlier and which lined the waterfront, were seized upon by property developers who saw profits to be made. Safeguarding the wharves has been a major and contested issue for the PLA. There are 70 wharves remaining on the River Thames, following the loss of many viable working wharves in the 1980s.

The Port of London is a different animal today, which has shifted to a model of riverside wharves and a port that has moved eastwards more able to respond to the dynamics involved in international shipping, where the imperative is to get the ships in and out as quickly as possible. As Martin put it:

> What is the Port of London? It's simply a generic name for all of the different port facilities on the River Thames. Where is the Port of London? And the answer is- all along the Tidal Thames.
>
> By 2017, the Port of London Authority's role had changed to acting primarily as a navigation and pilotage authority, with powers of regulation and control over activities on the river, lending itself to a comparison with the Highways Agency and their responsibility for the repair and maintenance of road. A digitalized operations hub at Gravesend charts the movement of every craft on the river, following the arrival of ships at the estuary and their movement up the Thames through transponders and radar. One of their major responsibilities is to keep the river safe and clear, increasingly deploying new sonar technologies to detect rubbish and sunken objects or spills on the river bed, and installing lights on the river, after a tragic accident in the 1989 resulted in the death of 51 people on the Marchioness party boat which sank after being pushed under by the dredger Bowbell.

Following years of relative inactivity on the river, the late twentieth century heralded a revival of the Thames as a throbbing artery in the body of the city once again. On the waterfront new initiatives flourished with the opening of the Tate Modern in 2000, and other cultural activities spread along the South bank with the construction of the new Globe theatre in 1997, while numerous festivals and events spilled over the banks. The river itself enrolled new players on its waters most significantly with the explosion of barge traffic from the inner city to the estuary, primarily to shift the tunnelling soil from Crossrail (the new high-frequency, high-capacity railway for London and the South East) to reconfigure the empty marshlands as a bird sanctuary. Another major infrastructure project, the Thames Tideway project, the largest sewage system to be built in the capital since the Bazelgette system was constructed following the 'Great Stink' of 1858—when the hot weather exacerbated the smell of untreated human waste and industrial effluent—is on track to follow. As Martin put it:

> I've heard people say that probably could not have been achieved in a modern, busy, congested capital without using the river, because if you think about it every ton of that waste would have to have gone onto a lorry and then driven out of London—you'd still be building it. They've built a huge new bird sanctuary in Essex with a lot of the debris… 'cause it's soil and chalk and so on.

Vast quantities of waste are also removed from four wharves in the along the river at Western Riverside Waste in Wandsworth, Smugglers Way in Vauxhall, Northumberland Wharf in Tower Hamlets, and one in the heart of the capital—Walbrook Wharf—alleviating congestion on already busy roads.

James Trimmer, Director of Planning and Environment for the Port of London Authority, explained the process:

> So the rubbish is being collected from the doorstep, it's going into the centre, it's being recycled, it's being sifted out and then it's being loaded into the containers, the containers are going into the barge and then away they go. You need the reversal of that for product, you need a logistics centre, a big logistics centre that takes in the containers, breaks them down small enough to go off into the vans and smaller lorries and distributed. And you haven't got the space and you can't create that any more.

As James pointed out, for many months of the year, the river is nevertheless an underutilized asset in the heart of the city 'a six-lane motorway going all the way from Putney to the Thames Barrier'. Yet, gradually, a renewed vision of the Thames as vibrant water is emerging, accompanied by its new vitality as a space of circulation, pleasure and recreation, as we see shortly.

River Crossings

Bridges mobilize passions, captured most poignantly in 1802 in Wordsworth's famous poem From Westminster Bridge:

> Earth has not anything to show more fair:
> Dull would he be of soul who could pass by
> A sight so touching in its majesty:
> This City now doth, like a garment, wear
> The beauty of the morning; silent, bare,
> Ships, towers, domes, theatres, and temples lie
> Open unto the fields, and to the sky;
> All bright and glittering in the smokeless air.
> Never did sun more beautifully steep
> In his first splendour, valley, rock, or hill;
> Ne'er saw I, never felt, a calm so deep!
> The river glideth at his own sweet will:
> Dear God! the very houses seem asleep;
> And all that mighty heart is lying still!

Bridges represent a space of transition and ambivalence. There is something compelling about standing on a bridge over the Thames watching the tides swirl below. On the darker side, a dozen or so people each year throw themselves off a bridge in London, and rarely survive. So too Westminster bridge became the site of a terrorist attack in the spring of 2017 killing five people and injuring many more.

Timber bridges across the Thames were constructed, and re-built as they rotted, from Roman times. Richmond Bridge is the oldest bridge crossing the river built from stone between 1774 and 1777 and funded by tolls, which was constructed as a replacement for a ferry crossing connecting Richmond town centre on the east bank with its neighbouring district of East Twickenham to the west. The last bridge to be completed across the Thames was the Millennium Bridge connecting St Paul's to the Tate Modern and initially a site of contestation and panic, as its original design left it vulnerable to swaying in the winds.

Material and human connections are nowhere more apparent on the river than in the construction of bridges, as the watermen were quick to see. Two stories from two London bridges tell different stories which illustrate how the very matter of bridges embeds and constitutes different publics between the banks. On a journey down the Thames on a tour boat, you might well hear the operators refer in passing to Waterloo Bridge (originally built in 1811–1817 and designed by John Rennie) as Ladies' Bridge. This is a hidden history[3] illustrating one aspect of women's work during the war when in 1941 women were called upon by Ernest Bevin, the Minister of Labour and National Service, to alleviate an acute manpower crisis on the home front, and to work in the heavy industries, such as shipbuilding, engineering, construction, aircraft manufacture and munitions.[4] As war ended their contribution was quickly forgotten and women were cajoled back into domestic life. The story of the women construction workers on Waterloo Bridge was unearthed from the archives in the early 2000s by historian Christine Wall, who was joined by film-maker Karen Livesey to pursue the story through oral history. *The Ladies' Bridge* documentary explores why these bridge builders and 25,000 other female construction workers had been written out of history. As the son of a crane driver on the bridge explained in the film:

> I can remember seeing ladies here… I think there was quite a few hundred ladies up here they did the less technical jobs, the lifting and the tugging where the men done the crane work and the technical type of work. And the ladies were in two grades of ladies. The ladies with the turbans and the dungarees you know with the bib up—there was more of them—but the ladies which were like the senior lady that could drive and could undertake

---

[3] Sheila Rowbotham.
[4] So beautifully captured in the film *Rosie the Riveter*.

a more of a technical job like, they wore an all in one overall a bit similar to the men…They probably didn't remember the women working on the bridge because they didn't look like women, you know if you have a flat cap on and an overall all in one, even today I have a lot of trouble to see who's a man and who's a lady. My father he used to love the ladies.

So great was the ignorance (or disavowal) of their contribution to the reconstruction of the bridge, that in 1945 Herbert Morrison—the then leader of the London County Council—on opening the bridge praised "The men who built Waterloo Bridge are fortunate men. They know that, although their names may be forgotten, their work will be a pride and use to London for many generations to come. To the hundreds of workers in stone, in steel, in timber, in concrete the new bridge is a monument to their skill and craftsmanship."

The Ladies' Bridge film, screened first in 2006, has raised the women's invisibility as a matter of concern and mobilized a multiplicity of publics, to press for the women's work on the bridge to be articulated and recognized. In 2009 the Women's Engineering group, ARUP and Women in Property started the Purple Boots and Decent Wear for Work campaign to raise the profile of women workers on the bridge. By 2015 Historic England had amended its history of Waterloo Bridge to include the participation of women in the work force, and in 2016 the bridge was lit up with a human chain crossing the bridge in celebration. The bridge of women had won its battle.

Tower Bridge tells a different story, one which foregrounds technology, tourism and a different kind of involvement of multiple publics. This is a bridge which stands in for London, arguably its most potent icon represented in films, objects and postcards and visited by thousands of tourists a year. This is a combined bascule and suspension bridge built in in 1886–1894 and copied from a Dutch design. It is one of five of the London bridges owned and maintained by a charitable trust—Bridge House Estates and overseen by the Corporation of London. The bridge consists of two towers joined by two horizontal walkways, designed to withstand the horizontal tension forces exerted by the suspended sections of the bridge on the landward sides of the towers. The objective of its innovative design was to enable the access of sailing ships along the

4 River Powers: Assembling Publics, Connections and Materials…

Thames at the same time as allowing traffic to cross the river through the capacity to raise the lower level.

Peter, a senior technician at Tower Bridge, described its workings:

> Bascule—It's French for seesaw. And a bascule bridge, you've got the long lever, which is the road, and on the back end you've got the short part which is counterweight, and it's balanced, so that's 12 tons. So when we're talking about two bridge drivers and separate bascules, the bascules in those days were operated independently so you had two bridge drivers, one north, one south, and in those days you would have always seen the south side bascule rising first, so they'd have always been out of step, 'cause basically the south side bridge driver had control of the nose bolts which locked the two halves together. So the north side driver couldn't start lifting until he saw the south side go, because he wouldn't know whether the bridge was unlocked or not. So that's why that was.

The original mechanism was based on water hydraulics developed in the nineteenth century, which in 1974 was replaced by a new electro-hydraulic drive system to raise the lower level bridge. Digitalization of the bridge in 2000 enabled the installation of a computer system (programmable logic control) to control the raising and lowering of the bascules remotely. Three senior technical officers and six (so-called) bridge drivers operate the bridge during the day and night-time.

Bridge statistics recording bridge lifts provide a lens to the wider sociocultural and economic context. When Peter started working on the bridge in 1994 there were 400 bridge lifts annually, a number which had decreased from thousands of openings in the post-war period over many years following the closure of the docks. Since then there has been an annual increase to almost 1000 lifts in 2010.

Peter explained why:

> The amount of pleasure craft on the river, you've got all the cruise liners coming up, war ships coming up, and during the summer months you have a lot of sailing barges up and down, they do trips up and down. ….Some of them are privately owned and some are companies like Tate & Lyle and people like that, and Allied Mills, they've got their own sailing barge and they use it for corporate clients and trips up and down the river, and of course a big part of that is coming through Tower Bridge.

This then is a story of the river playing a growing part in global corporate life, new forms of tourism and pleasure craft and leisure practices. Peter:

> Have you seen the Dixie Queen moored down there by Butlers Wharf? She's a corporate vessel that does a lot of events, they sent it as a package, and they made their stacks on top of their vessel just high enough to warrant a bridge lift! And of course part of the package for all their clients is to see Tower Bridge lift as it goes up and down the river. So a lot of bridge lifts are done for her, especially weekends and during the evening. So that was a significant increase, for her. But it's all tourism.

An exhibition space built in the twin towers in 1982 further attracts tourists to the site.

Rebirth of the Thames: Connecting Places

In the boroughs near the river, the haunting sound of the foghorns at night until the end of the 1950s suggested the mysterious presence of barges wending their way blindly through the thick mists gathering above the water's surface. By the late 1950s the foghorns had gone with the enactment of the Clean Air Act bringing an end to the thick fogs that had smothered London for centuries. Three decades later the constant refrain amongst planners and policy makers was that the potential of the river was not being recognized. By the 1990s, after years of the city turning its back on the river, a different era was born with the emergence of new forms of circulation and cultural initiatives enrolling new subjects on, and besides, the river.

In 1877 Thomson lamented the passing of the Thames as a major thoroughfare. Early passenger transport had been provided by the 'Western Barges', known for their discomfort with only a bundle of straw to sit on, followed in the seventeenth century by the arrival of the steamboats, which by the nineteenth century had displaced all the old barges and wherries. The London Omnibus and Thames Steamboat Guide (1851) details the daily timetable of eight steamboats that ran from Hampton

4 River Powers: Assembling Publics, Connections and Materials...

Court to Gravesend with such colourful names as the Ant, Bee and Sunbeam Steamboat Company and the Diamond Steam packet Company (with boats named after jewels displaying checkered white diamonds on their black chimneys) each leaving every 10 or 15 minutes from 8 am to dusk at a cost of between one halfpenny and six pennies depending on the distance.

The birth and growth of the Thames Clippers to a record high of 4.3 million passengers by 2016 is a story which revolves around a passionately committed individual, enabled by new shipping technology, which in turn has constituted the river as a space of circulation, after many years that had witnessed the long decline of the passenger services as rail and omnibus services, followed by the rise of the motor car on London's streets, had gradually diminished their significance. This is a site of entanglements between individuals, habits, regulations, institutions, technologies and everyday cultures. Sean Collins, a waterman born in Bermondsey and steeped in the cultures and histories of the river, launched the concept in 1999. A third-generation waterman, his father started life as a docker and shifted to lightering as employment in the docks receded. Sean's two cousins work on the river as managers of another passenger boat service as did their father who was a lighterman for Cory Environmental, a large resource management, recycling and energy recovery company. Collins articulated fond memories of travelling with his father on the barges from Tilbury to Brentford as a small boy. Like many of the watermen Sean had a passion for rowing, winning the Doggett's race in 1990 and competing in the Henley boat race 13 or 14 times. This is a culture which still permeates boundaries between work and leisure, creating passionate allegiances and daily rituals. Sean is one of the watermen appointed to the Queen where they sport a dashing red coat and accompany the Queen on her royal barge for ceremonial occasions (Photo 4.3).

Sean started his employed life on the river as an apprentice on ThamesLine (where he rapidly rose to captain), which became the River Bus Partnership—a river bus service that was created in 1985 under the Business Expansion Scheme introduced by the Thatcher government for developing enterprise areas and the docklands. This initiative quickly ran into financial difficulty, and was taken over by the Reichman Brothers of

Photo 4.3 The waterman's coat. Sophie Watson

Olympia and York to ensure that a continuation of the service to Canary Wharf in its early days, alongside the DLR. Further financial difficulties led to renewed management under P&O for the Canary Wharf Group. Here another individual (a small shareholder at P and O), Victor Hwang, the owner of Battersea Power Station, emerged as a key player with a vision for the reconstruction of Battersea Power Station where the idea of a river bus service to increase accessibility to the city centre coincided with Collins's interests. Hwang owned a fast ferry builder on the Isle of Wight called FBM and was producing fast ferries there that were being exported to Hong Kong, for the Hong Kong to Macau route, which is where he started. This then is a story of multinational and global capital, where the Thames has enrolled wealthy investors from overseas and mobilized new international connections.

For Sean Collins ThamesLine had been

> an amazing experience to see how a river bus service could potentially pan out and what impact it could have on a city. ... so I was very disappointed

4 River Powers: Assembling Publics, Connections and Materials… 95

when the administrators finally pulled the plug on the River Bus Partnership. That was '92. And there was three new boats that were built by the partnership at FBM …and they was the sort of next stage, the next era and far, far better boats than the first eight that were built by ThamesLine. The right craft, at the right time, managed properly, I could see that there was the opportunity there to make a success of it, but it was a good few years until I actually done it. For two reasons, one I became quite despondent in the river, the wages had really gone down, there was very little work.

For some years following the company's demise, Collins withdrew from river transport to concentrate on his rowing. Central to this episode once again were the watermen's strong interconnected networks, as Collins became involved in training a young man for the Doggett's race, whose father Alan Woods owned Silver Fleet—the premier charter boat company on the Thames for parties and corporate events at the time. A conversation developed between the men concerning the potential for a riverboat service, where Collins expressed his sadness about its failure. As Sean described

> Then ironically we started looking at something for the millennium, to bid jointly with Battersea, where Alan was going to expand on his leisure and tourist boats and there was the opportunity for a fast ferry service as well. And we were unsuccessful with that, but that said, there was another attempt, I think about the twelfth or thirteenth attempt, with White Horse Fast Ferries who built, I think, six boats and won the contract through TFL to run a service for the millennium alongside the slower tourism service, sightseeing service, City Cruises.
> In between that time of that actually coming to fruition and when the bid closed, one of the old river buses came on the market. Alan and I spoke about it and he said, 'Look, if you're interested in this I'd be very keen to go with you on it.' And Alan was somebody that I'd always aspired to because he always invested his money back in the river. And that's what we done. We got the first boat, we started in 1999 with one boat.

Seen as a competitor to White Horse Ferries, the new venture was not permitted to stop at any of the London river services, TFL piers, thus restricting their access to the private piers—Greenham Pier, Canary

Wharf, London Bridge and Savoy. Unmet demand led, from the very first week of the service, to the boats bursting at the seams with commuters on the initial morning and evening services. Ironically the day the pair gained ownership of the second boat in Plymouth, they received a call offering them a craft from the White Horse Ferries which has just ceased trading. As Sean put it: 'so I'd got two boats for the price of one that day!'

Cross-river services rapidly expanded between the businesses at Surrey Quays, Canary Wharf and Masthouse Terrace on the Isle of Dogs. The establishment of a contract followed with an Australian ship builder that had built boats for Sydney Harbour, who suggested a deal for the development of the Hurricane Clipper—the first of their 220-seater boats.

Sean again: 'And that created an amazing wow factor, it gave us the carrying capacity on those main runs in the morning and the main runs in the evening to deliver a service and not disappoint anybody, because we was constantly leaving people behind. And it grew from there.'

Clippers went from strength to strength with a subsidy from Greenwich Council and Transport for London under Mayor Ken Livingstone's office. In 2005 two further boats were added. With the demise of White Horse, the Clippers expanded their operations to and from some of Central London piers, including the London Eye, and the two Tate galleries. By 2008 they were carrying over a million passengers a year.

A different set of entanglements emerged between the Clippers and property developers where mutual interests coincided. The Royal Arsenal at Woolwich, a former site of armament manufacture and testing employing close to 80,000 people at the height of the war, had finally closed its factory in 1967 leaving a remote and isolated tract of land. Over 40 years later, with the expansion of riverside development, a property developer Berkley Home saw profits to be made from this disused and neglected tract of land, yet commercial success was dependent on fast transport services to the city.

The extension of Clipper services to the site was the obvious solution. What we see here are the entanglements of the personal, the cultural—the history of Watermen and their work, the global/financial and the administrative, which together have created this new emerging public.

This brief account of the rise of the Clippers follows a trend in many global cities. New York City's mayor, for example, as of 2017 invested 380 million into a fund to expand river services or port services up both the Hudson and the East rivers. Having started late, London is regarded as leading edge, with its expanding services, new technologies, connections with business, its role in education—where there is an emphasis on children as the future, and its forward thinking. Sean is one of the leading figures in the international network of ferry operators—Interferry—his life now a far cry from the life of his forebears as watermen and lightermen on the Thames. It is a story which foregrounds the role of water in actively remaking place, where interconnections between transport, housing, employment and leisure mutually reinforce the vibrancy and viability of each of these urban components.

Watery Pleasures

We have seen how the decline of the river as space of work deriving from containerization technologies, shifted everyday cultural and economic practices in dramatic ways. In their place, the river has increasingly enrolled new subjects and actors constituting multiple publics in practices defined by leisure and historical concerns. Some of these practices are continuous with former traditions but have been transformed and reconstituted with new meanings and rationalities. Digital media and new technologies have been central here.

Mudlarking in its contemporary form offers an illuminating window into processes of socio-cultural, economic and technological change on the river. Gone are the impoverished women scraping a meagre living on the banks of the Thames. In their place a new breed of mudlarks has arisen mediated by the emergence of a more widespread interest in historical and archaeological concerns and often driven by affect and passion. This is often an individual pursuit, motivated by a quest for solitude and passion for old objects inflected with traces of the past, and redolent of another era. Nicola, who gave up investment banking to pursue her life as an artist and mudlark gave an illuminating account of her practice and its resonances with other lives:

I find so many fragments of lives down by the river, you know, little objects. I mean, these three things I wear round my neck, they're all individually found items. There's a tiny cross from the 17th century, the Georgian heart, which I had soldered onto a piece of silver 'cause it was so delicate and a Victorian heart charm from a bracelet.

The luggage label, it's just a very small, innocuous-looking little piece of brass and initially when I found it, I thought it was just something off of a… like a shop name or something. But in fact, what it was was a name and an address and all it said was 'F Jury, 72 Woolwich Road'. And because there was this name and address on it, through all of the sites on the internet and with some help from people on Twitter, we were able to trace his entire life and I actually did write a blog about it. His name was Frederick Jury and he was a World War I soldier. He actually went to enlist in Australia in 1916 and he was 42 and of course there was the question about why would he go over there and not do it here? And it's probably 'cause he might have been too old to enlist here. So he went over there and then he came back and he fought in the trenches in Belgium and France, he was hit by a grenade. …So when I found this little piece of metal, it was like opening up a storybook of somebody's life.

….So that's one of the reasons why I'm fascinated with mudlarking 'cause each thing I find is about somebody. I mean, who gave somebody this heart? Why is it in the river? What was their life like, was it a sad story, is it a happy story? So that's why, for me, that's what keeps me going back.

And

I'm a solitary mudlarker, really, because I find it very meditative and I'm always very much in the moment when I'm mudlarking.

Every visit to the Thames reveals new objects, whether a piece of glass and pottery, or buttons and coins, or more potentially valuable items like Tudor buckles or jewellery. Where objects are thought to be of historical significance—namely over 300 years old, mudlarks are required to take them to the Museum of London to be declared under the Portable Antiquities Scheme, where a picture is being assembled of the life of the river.

According to Nicola there is an obsessive element about the pursuit, which attracts loners and predominantly men, though the demographic

4 River Powers: Assembling Publics, Connections and Materials… 99

is apparently changing following the BBC series—Mudmen, featuring one of the most well-known mudlarks—Steve Brooker. New publics have emerged enabled by social media and digital technologies. Nicola for example has an Instagram account, a website, a YouTube channel and a Twitter account to engage people in what she has found. Instagram provides a device to enable mudlarks to help each other identify objects, in this sense mobilizing a virtual community of knowledgeable practitioners. Other mudlarks attend the monthly meeting of Thames Field where new discoveries are revealed.

As Nicola put it:

> [T]hey love the objects, I think. I mean a lot of them talk about when you find a coin, it's almost like somebody's handed it down to you from the 16th century or something like that.
>
> To think that the last person who touched the Queen Elizabeth coin could have been somebody… and it's like it's been handed down to you. It's history you can touch, it brings it alive. It's good for kids as well. I do genuinely think that for most of them, it's that love of the objects. And you know there's a little bit of competition as well. It's like, 'Oh wow, I've found this. I found this pilgrim's badge from the 17th century'. And then people quite enjoy taking pictures of all the things they've found and showing people. You know, it's a little bit of, 'Look at this'. Like the Thames and Field, they meet once a month and they have a table that everyone puts their favourite find and then everyone votes on the best thing.
>
> It's a bit like Show and Tell. Everyone brings their little thing to show and they talk about how to clean coins and various little things that are going on…. we all like to think about the history of these things. I mean even a button, it's often got a name of a tailors on it.

Digital technologies also enable mudlarks to choose the right time of day to dig, through the tide app on their phone.

Mudlarks distinguish themselves in their practices with imagined hierarchies of authenticity. Nicola again: I work on an 'eyes only' basis, I've never used a metal detector and there are reasons for that; I like the organic way and when things come to the surface. I've been in places where there are metal-detectorists and they'll miss something perfectly lovely right in front of them.

There are also practices of 'appropriate' conduct where intimate objects are concerned such as messages in a bottle saying goodbye to loved ones or revealing secrets, and many of the mudlarks consider these should be returned to the water. Formerly mudlarking required no permission, but under new regulations issued by the Port of London Authority mudlarks are given permits that define the depth to which scraping is allowed. An interest in the history of the river has mobilized other such groups and enrolled other constituencies. The Thames Discovery groups run a scheme called the Foreshore Recording and Observation Group—the so-called FROGs, where training is provided and members record discarded objects on the river banks, and offer guided walks on the foreshore.

Fashionable Pleasures

Fifty years ago, the river was little more than a public drain. By the end of the decade of the twenty-first century, following initiatives driven by the Environment Agency and water authorities, in tandem with the Port of London Authority, reports of salmon and otter returning to the Thames after centuries of filth deterred most living creatures, filled the press (McCarthy 2010). The cleaning up of the river has had unintended effects—and that is the growth of river sports typically associated with Sydney harbour or with a clement climate. Improved technologies including wetsuits and boat craft, combined with growing incomes in certain sectors (notably finance and media) of the capital's population, whose holidays in exotic locations have inspired a taste for water sports, have brought a plethora of activities to the river, from canoeing to kayaking, paddle boarding to jet skiing. The Port of London Authority reported the many challenges this has brought.

As Martin described:

> A lot of it is touchy feely and compromise and give and take, but that's what we have to do. And then with the motor boaters we've brought in speed limits, which are controversial, but …there was a case recently… I'll send you a picture of his boat. And he zoomed through London, where we have a 12 Knot speed limit, 40 or 50 knots, I mean really reckless.

4 River Powers: Assembling Publics, Connections and Materials…

As James from the PLA explained, the Thames as a tidal waterway is subject to a common law right of navigation where no license is required to travel or drive a vessel—you could even go in a bath tub subject to it complying with our bylaws, general directions and the international code of the sea.

This free movement only changes at the point where the river becomes fluvial where vessels are required to be licensed. As James put it:

> I'll denigratingly refer to gin palaces; one could get on a… drive up through central London, through the busiest inland waterway and probably the most dangerous in terms of driving up with the bridges, without, if you do it yourself, without any consent whatsoever.

Rowing has always been a feature of life on the Thames, though primarily for transport rather than leisure. Over the years, amateur rowing has become popular as schools and colleges send teams of students to train on a daily basis. An early morning walker along the riverbank could not help but be distracted by the elegant rowing boats and skiffs swishing by. Over the centuries, angling too was a part of life on the river, often for food, but also for pleasure.

Sailing Barges

Traces of the past live on in another current activity which has attracted devotees. Like the mudlarks, passions for objects of the past have translated into new cultural practices in the revival of the sailing barges. Pictures of the Thames from the Elizabethan era depict these striking structures with their massive and billowing sails wending their way through the Thames waters. The Thames sailing barge was a structure specific to the river, which developed over the years with diverse influences particularly from the Dutch barges which sailed around the lowlands. But the Thames barge was unique with its rig that could be handled by one man, unlike the old trade boats which had large square sails and a high mast, necessitating a crew of 10–20 sailors, which the barge owners could not afford.

There were some barge companies such as the London and Rochester Trading Company and some of the mills which had fleets of barges, and some were freelance where men raced with their barge for a contract or job rushing to be the first to the site and banging on the door for a cargo home. Barges on the Thames were used for agricultural products—grain and straw and sometimes timber. Peter, a master barge master gave an illuminating account:

> They might do what they call 'the rough stuff' which would be the rubbish from the streets of London, the house muck and the straw and stuff. They'd pick that up in London and take it out to the brickworks in Kent, and the brickworks would put it into the clay mix and when they fired it burnt in the brick, which if you look in London on a red brick building you'll find flecks of black, sort of ash in the brick and that's what that is, it's the rough stuff that they took out of London, put into the mix, fired the brick, it enhances the brick, it heats it from the inside, it makes it stronger, makes it fire better. And the bricks would be then transported back on the barges back into London from the brickworks.

These barges had flat bottoms which meant they could enter a muddy creek where mud and clay could be assembled by large spades and thrown straight into the empty barge. This was a tough and arduous existence where bargemen were considered to be of a far lower status than the watermen and lightermen; 'they were kind of the delivery truck drivers of their day, and they were looked down on a little bit by the big ships and the dockers, they saw them as a bit of a nuisance really, just doing of the job'.

After the war engines were introduced to enable easier manoeuvres in the docks, since a lack of control in the crowded space was hazardous. The last active sailing barge on the Thames ceased working in 1972. Following the trends of reconstituting old traditions in new leisure practices and evoking a reworking of history, new associations have emerged which bring together new publics. There are currently 35 active sailing barges, and the Thames Sailing Barge Trust is one such association which owns a couple of barges—the Centaur an 1895 built by Canns of Harwich, and Pudge built in 1922 by London and Rochester Trading Company, and part of the Dunkirk landing. Participants are assembled in organizations like the Association of Bargemen, which resembles a union, to restore the craft, preserve the skills and train people to become sailing barge skippers, 'just like the old guys were'. Gone are the days of transporting cargo, now the barges can be seen on the Thames carrying corporate parties for an outing, or disadvantaged children from inner London. Safety requirements which are now so central to everyday life in the 'risk society' mean that new navigation systems are required where a transponder enables the barges to be seen from the PLA control centre at Gravesend.

As Peter from Thames Sailing Barges put it:

> [W]e're not very maneuverable like the Clippers are and we're old and a bit cranky so the Thames is not a friendly place for us anymore.

Not surprisingly this is a gendered pastime, with mostly male skippers, although weekend working parties attract more diverse publics, including women who have particular interests such as painting the boats. Like the mudlarks this is an active social media world, where networks share information about upcoming events and meetings.

Conclusion

This exploration of the Thames and its powers is one which bears many resonances with cities all over the world. If we imagine Paris, we think of the Seine; if we imagine Rome, we think of the Tiber; and if we imagine New York, we think of the Hudson. Rivers are iconic. Rivers, as we have seen, have the capacity to enable commerce and trade, to enrol multiple publics, and to connect people, places, materials and technologies. Their very materiality, the flow of the water, has its effects. Rivers are vibrant and powerful and have the capacity to drown people in their waters, at the same time as enabling everyday pleasures—meandering along the river banks or sitting on a boat as it glides through the city. What we have also seen is how the meaning of rivers shifts and changes over time, from an unnoticed artery through the city to the motor for its economies and cultures. Across many parts of the world, we are seeing the revival of waterfronts as expensive new developments attract wealthy residents and overseas investment, contributing to the growing inequalities in the city between the rich and the poor. Where once industrial sites adjacent to the river were left vacant and forlorn, run down warehouses are transformed into new cultural spaces and engaging new publics and new economies of consumption. In so many ways, rivers in cities matter.

References

Coates, Peter. 2013. *A Story of Six Rivers History Culture and Ecology*. Reaktion Books.
Dickens, Charles. 1865. *Our Mutual Friend*. Chapman and Hall.
Gosling, Harry. 1927. *Up and Down Stream*. Nottingham: Spokesman.
http://www.dailymail.co.uk/news/article-2364712/The-forgotten-heroines-built-Waterloo-Bridge-Historian-reveals-women-drafted-construct-The-Ladies-Bridge-WWII-got-credit-deserved.html.
Mayhew, Henry. 1861/1986. *The Illustrated Mayhew's London*. Edited by John Canning. London: Weidenfeld and Nicholson.
McCarthy, M. 2010. Britain's Rivers Come Back to Life. *Mail*, Friday, December 31, 2010.
Steadman Jones, Gareth. 1971. *Outcast London: A Study in the Relationship between Classes in Victorian Society*. London: Clarendon Press.
Stow, John. 1876. *A Survey of London: Written in the Year 1598*. London: Chatto and Windus.
The London Omnibus and Thames Steamboat Guide. 1851. Published for the Proprietors by S. Johnson, The Strand.
Thomson, John. 1877/1994. *Victorian London Street Life*. New York: Dover Publications. Originally published 1877. London: S. Low Marston, Searle and Rivington.

5

Embodied Water Entanglements: Sex/Gender, Race/Ethnicity and Class Urban Sanitation Practices

Bodies, cities and water are mutually entangled in a myriad of ways. Not only do bodies need water to survive, they also need water for washing, sanitation and maintaining cleanliness. The engagement of bodies with water represents a key site of everyday material politics, which is constituted through material artefacts and technologies, urban regimes of government, the availability of water, and institutional networks. The entanglements of bodies and city waters are not fixed in time or in place, but instead are the outcomes of complex local and national networks, economies and geographies, as well as cultural expectations, routines and habits. The rituals and customs associated with washing our clothes, and ourselves, or urinating and defecating are as different in London from 1850 to 2017, as they are currently different from Kolkata to New York, or the townships of South Africa to Kyoto. Washing and laundry practices in cities like London bear little comparison with cities of the Global South, where the luxury of plentiful water supply and the widespread current use of domestic washing machines is the privilege of but a few. Here washing is largely a public affair, where water is available at streams, rivers, wells and pumps, with contrasting configurations of public/private and gender relations. What people take to be normal in one place may be completely different in another.

© The Author(s) 2019
S. Watson, *City Water Matters*, https://doi.org/10.1007/978-981-13-7892-8_5

As Taylor and Trentmann (2011) suggest, water plays a part in a larger history of politics than everyday but each is entwined with the other: 'In conflicts over water rates, baths and drought, the politics of everyday life contributed to the shift from liberalism to a social democratic politics of provision. Any study that focuses on the quotidian must, by definition, recognize the highly localized nature of practices and conflicts.' As Taylor and Trentmann suggest (2011: 141) how people do their washing, eat or perform other practices should, therefore, move to the centre of analysis, since it is as important as what engineers, social reformers and economists have thought about how people should behave. Most significant here is the differentiated nature of bodies, where sex/gender, race/ethnicity and class constitute, and are constituted by, practices of keeping oneself and one's body clean (Skeggs 1997).

Watery embodied entanglements take a material form. Yet mundane domestic objects—particularly those associated with the body—which in various ways and at various times move out of the home and animate an assemblage of multiple spatial forms and socialities in the city go largely unnoticed (Shove's [2003] work is a notable exception). One explanation for this lies in our sense of disgust at body effluent and waste, the 'simple logic of excluding filth' or expelling things that are seen as abject (Kristeva 1982), or our need to exclude uncleanliness to maintain boundaries (Douglas 1988: 41), or in an implicitly racialized notion of dirt as dangerous (Sibley 1995). My argument here is that the very invisibility of these processes also lies in their sexed and gendered nature. They also constitute multiple publics and public spaces as we shall see.

Washing Bodies

What one culture deems to be appropriate levels of bodily cleanliness another may see as excessive or wanting. Everyday washing practices differ across time and place, and articulate wider relations between private and public spheres as well as complex and contested norms around bodily practices, which are deeply implicated in sex/gender, class and ethnicity. Daily or twice daily showers, which characterize everyday life in many of the richer countries in the world, to a nineteenth-century urbanite would seem extraordinary, just as such pleasures and possibilities are denied to people living in the marginal settlements in many cities of the world. The provision of baths and washhouses, like the provision of public toilets, has not only reconfigured private matters as public, it has also generated new forms of politics and contestation. Interventions to regulate conduct around washing behaviours have constituted an important aspect of government attempts to manage populations, particularly those seen as inadequate, immoral or 'dirty' in some way.

In the UK, parts of Europe and the US, the provision of public baths entered the public arena from the middle of the nineteenth century, though even between these countries there were differences. The British notoriously gave little thought to bathing. The anonymous author of a short article in the *British Medical Journal* in 1911 wrote: 'How near cleanliness lies to godliness and all the virtues is a thought that rarely strikes the unobservant Britisher, who is apt to take his daily tub, his weekly bath or in extreme cases his yearly bar of soap and the sea, as a matter of course,' Not so the Germans where a one Dr Berger was advocating public baths in every town and village, to be connected to every public elementary school, hospital and factory. Goodwin Brown (1893: 144) writing of the people of America in 1893 wrote despairingly of his countrymen who as far as he was concerned were so far behind 'in one of the first essentials of a highly developed civilisation'. Making an interesting link between cultural practices and the materialities of everyday life he suggested that 'people who are unclean see no necessity for clean streets, dwellings and public buildings'. Many advocates for public baths at the time spoke admiringly of the Ancient World—in Egypt, Rome and Greece—extolling the magnificence of public baths enjoyed by the urban (often male) citizens. Here the advance state of hydraulic engineering in cities like Rome, even in Cicero's time, meant public baths were a ubiquitous feature of the urban environment.

The first public baths in Britain were established in Liverpool in 1828 at St George's Pier and were salt water, followed 14 years later by freshwater baths on Frederick Street (1842). The movement for public baths and washhouses was practically initiated at an influential meeting at the Mansion House, under the presidency of the Lord Mayor, in September 1844; when resolutions were passed for the formation of an 'Association for Promoting Cleanliness Amongst the Poor'; and a system for subscription was introduced. The first 'Free Baths and Wash-houses' was opened in a poor area near the London Docks, in Glasshouse Yard, Rosemary Lane, in an old but spacious building which had been occupied by the poor and homeless for sleeping. At minimal cost the building was converted to house a supply of tubs, boilers and several baths, where soap and soda and hot and cold water were provided for free. The intervention was clearly motivated by the notion amongst philanthropists at the time that the poor needed to be made clean:

> Medical men, clergymen, city missionaries, parochial officers, and all whom either professional duty or benevolence had led to enter the dwellings of the very poor, however their opinion differed in other respects, were at least unanimous in declaring that those dwellings exhibited a degree of dirt and squalor with which health and morality were alike incompatible… It had been allowed by all who were really acquainted with the homes of the very poor, that in their crowded and wretched dwellings cleanliness was impossible. In such places not only were there scarcely the means for personal cleanliness, but to wash and dry clothes properly was quite impracticable. It was proposed, therefore, to see whether the establishment of places where, for a small charge, a warm bath could at any time be had, and where all the conveniences for washing and drying clothes should be provided free of charge, or at a trifling cost per hour, would not be gladly accepted by the classes most requiring such conveniences. (VictorianLondon.org)

The establishment of baths began in earnest in the middle of the nineteenth century following the Baths and Washhouses Act of 1946 which encouraged local authorities in their construction. Crowded living conditions and the prevalence of disease was a motivating force, strongly supported by the medical profession, to improve public health and hygiene. In 1847 public baths were constructed in Gouldston Square (and finally demolished in 1989 to make way for the Women's Library), followed over the next decade amongst others by the construction of baths in Marylebone, Westminster, Poplar, Bloomsbury, Bermondsey and Hanover Square. Descriptions of the baths can be found in the Second Supplement to the Penny Cyclopaedia of the Society for the Diffusion of Useful Knowledge 1858 (Allsop 1894; VictorianLondon.org):

[A] bath-room is a distinct compartment, somewhat more than six feet square, shut in by walls of painted slate, which are carried up to the height of some ten feet: but the top is open, so as, while insuring privacy, to admit of thorough ventilation. The bath, in some establishments sunk in the ground, in others placed as usual above it, is either of iron enamelled, or of zinc. …On each door is a porcelain knob, having a number painted upon it; a similar number is painted inside. An index outside enables an attendant to let in either hot or cold water, as the bather may direct. The most perfect cleanliness is indeed observed in every respect. (VictorianLondon.org)

Like public toilets, these were gendered spaces which embodied normative assumptions about women, and involved different practices. The Act provided that women and men should bathe separately. In Goulston Square the baths for each were on the opposite side of the building, for example, and a walk past the St Pancras baths today, on the corner of Grafton Road, reveals the (newly embossed in gold) separate quarters for the sexes. In some baths, such as East Smithfield use of the baths by men and women were determined by different times of the day. Women were found to use the public baths less than the men, which was attributed to their embarrassment at bathing publicly, the greater cost of the baths in proportion to their wages, their dislike of unclean baths and the demands on their time due to domestic matters (Campbell 1918: 35).

Class divisions were also materialized in these buildings, with separate entrances for the first- and second-class baths and a differing price structure was laid out in the Act. Thus 6 pence was charged for a first-class warm bath, 3 pence for a first-class cold bath, 2 pence for a second-class warm bath and 1 penny for a second-class cold bath. According to the Penny Cyclopaedia, 'The first- and second-class rooms are usually alike in every respect, except that the fittings in the first-class rooms are of a superior kind, and more complete than in the second…. The charge for a first-class warm bath is sixpence, for which two towels, 'flesh and hair brushes, and a comb are allowed. For a second-class bath the charge is only twopence, but only one towel is allowed, and the bather must provide his own comb and brushes. The baths are in all respects alike, the same quantity of water (in most places forty-five gallons, but at St. Martin's much more) is allowed, and the bath is invariably cleaned after each person.'

An article in the *BMJ* in 1883 exposed the recycling of water from the first to the second-class baths, which provoked the author to urge managers of the baths to 'announce the fluid which they supply, whenever it differs from pure water'. The baths were highly used. In St George in the East during the year of 1846, there were 27,662 bathers, 38,480 washers and 4522 ironers (Quekett 1888). By the mid-twentieth century baths and washhouses were to be found in most municipalities. Bathing habits and routines continued to be distinct across different cultures, immersed in different regulatory practices and norms, and with diverse notions of appropriate washing conduct, such that a well-worn joke has persisted into recent years amongst Americans and Australians that the English kept coal in the bath, rather than bathing in it—a practice that was associated with the early days of private baths, often located in the kitchen with a wooden lid on top.

Public Toilets

The provision, form and design of public toilets and sewerage systems reflect and embed particular socio-cultural and economic materialities of spaces at specific historical moments. These are not value-free infrastructures to meet, or not, corporeal needs—namely defecation, rather, they constitute particular social, sexual and gendered relations and are themselves constituted by how sex, gender, class, race, sexual orientation are conceived at any one moment. At the same time the availability of public toilets and the efficiency of sewerage systems are mediated by economic and fiscal constraints, where inadequate sanitation, sewerage systems and toilets characterize much of the provision in poorer countries.

A turning point in the history of sanitation in the UK, or rather its visible inadequacy, was the summer of 1858 when the filthy smell of the river Thames, clogged up with untreated human and industrial waste running directly into the river, reached such an extreme level that the Houses of Parliament were forced to close down. Many attributed the transmission of contagious diseases, and particularly three outbreaks of cholera to the dirty river, but the Great Stink, as it was denoted, provoked a level of concern that prompted action. The chief engineer of London's

Metropolitan Board of Works, Joseph William Bazalgette, came up with a plan to construct 82 miles of underground brick main sewers to intercept the sewage outflows and 1100 miles of street sewers to divert the raw sewage which flowed through the streets. The plan involved a number of major pumping stations, two of which—Abbey Mills in Stratford and Crossness in Erith—are listed for protection by English Heritage and have been transformed into cultural sites for present-day tourism. The system was opened in 1865 by the Prince of Wales, but not completed for another decade.

This remarkable system has survived until the present day but has been under increasing stress with the expansion of London and its population. As a result a vast new tunnelling project the approximate length of the Channel Tunnel, the Thames Tideway Tunnel—referred to in the media as the 'super sewer'—is currently underway. The tunnel has become a matter of concern in many localities where residents have objected to the construction of tunnelling sites in their locality, mobilizing in turn responses from the engineering company. Thus, in Southwark, Tideway's delivery manager articulated a desire to engage with communities and create a positive legacy around their sites. Interconnecting the worlds of engineering and art, artists were invited to apply for the chance to design a large-scale artwork to surround the work site for two years. This resulted in a giant creation 'Stories from the Sewer'—for which artist John Walter enlisted Riverside Primary school pupils—to adorn the hoardings around Tideway's Chambers Wharf site in Bermondsey. This massive infrastructure project will impose its presence for some years on the life of the Thames itself, as the soil and spoil are moved out of the city to the edge of London on a train of barges.

The history of public toilets in cities is deeply imbricated in prevailing assumptions, norms, prejudices and representations of sex/gender, and in some countries race (e.g. in the South of the US toilets were segregated between Whites and Coloured). They are also 'highly charged spaces, shaped by notions of propriety, hygiene and binary gender divisions' (Gershenson and Penner 2009: 1). *Ladies and Gents* (Gershenson and Penner 2009) provides a lively and fascinating history of public toilets in the UK and US. From the mid-nineteenth century public health concerns led to the construction of public toilets across London, which were for men; and made possible by the new sewerage system; the provision for women lagged far behind. Contesting the dominance of men's needs taking precedence in the public arena, The Ladies Sanitary Association strenuously campaigned for women's toilets from the 1850s—an issue that was later taken up by the Suffragette movement. In 1884 the Ladies Lavatory Company opened its own private public lavatories at Oxford Circus for women who travelled to the centre for a day's shopping (Greed 2009: 47).

Over several decades in the late nineteenth century, the LSA campaigned for toilets at sites where there was a high volume of traffic, such as in Camden town, meeting resistance from locals and omnibus owners on the grounds that they would lower property values (Gershenson and Penner 2009: 5). Penner (2001: 41) explains this resistance as deriving from the notion that women's toilets sanctioned their presence in the street violating the designation of middle-class women's place in the home, and turned women into public figures where their private bodily functions became visible and their sexuality was evoked. Such was the force of Victorian ideals of femininity and womanhood that women themselves were reluctant to use the facilities, laying the ground for the St Pancras Vestrymen to argue that were a toilet to be built women would not use it (Penner 2001: 45). In 1905, following decades of lobbying and resistance, the construction of women's conveniences in Glasgow, Paris and Nottingham, and a report from the Highways, Sewers and Public Works Committee, the Borough finally agreed to build a women's lavatory on Park Street at Camden Town, which remains on the site to this day (despite a period of closure during the 1980s) (Penner 2001: 47–8). By 1928 a survey revealed 233 toilet blocks in London for men and 184 for women (Cavanagh and Ware 1990: 17).

Following the Second World War there has been a steady decline in public toilets. In 1999 the British Toilets's association was established to address and campaign for their provision and to argue for the necessity of a cheap decent toilet for people travelling away from home for an extended length of time. Since the turn of the century more than 45% of UK's public toilets have closed down with the closure of all but one of Manchester City Council's 19 public toilets. With pressures on public expenditure they are not seen as a priority. Many facilities in public places such as the Royal Parks and railway stations have introduced charges. In

5 Embodied Water Entanglements: Sex/Gender, Race/Ethnicity... 113

this beleaguered context, which discourages some people, particular older people, from traveling away from home, a Community Toilet Scheme was established in Richmond in 2005, and followed by other London councils, which allows the public to use the toilet facilities in registered pubs, shops and cafes, which are subsidized by the council to cover costs for their involvement in the scheme. There are now approximately 70 premises participating in the scheme.

The shifting boundaries between public and private are no starker than in the trajectory of public toilets into private spaces. In London many of the old toilets have been converted into nail bars, trendy shops and even bars—such as *Ladies and Gents* in Kentish Town, where drinkers booze locally brewed gin in proximity to the former urinals. In South London a toilet was transformed into a luxury flat described by the estate agent as—'a snip at £125,000' in 2002 (Economist 2002). The scarcity of toilets has found its way into a digital representation the Great British Toilet Map which is a digital resource to locate the nearest toilet to you which meets your requirements. In the words of the director of the British Toilet Association: 'We are becoming a third-world country in toileting.' On this, of course he is wrong. The dearth of public and private toilets in many parts of the world is one of the devastating effects of living in poverty, where public infrastructure is minimal or falling apart.

On World Water Week in 2010 Water Aid drew attention to the limited progress made by the UN in developing its millennial development goal of better sanitation across the globe, with a large demonstration in Trafalgar Square using the slogan 'Dig Toilets Not Graves' erecting 167 shovels from the grass to symbolize the number of children who dies every hour in the 'developing' world from diarrhoea. In Mumbai (like so many other cities) the daily life of sanitation is precarious with, for example, the threat of the drains breaking down through storm water or the fear of children falling in to makeshift toilets, or getting washed away in high tide, as a regular occurrence (McFarlane 2008). In this context toilets are central to political life, where city dwellers are regularly confronted by changing tariffs, demolition and a broken-down infrastructure. As McFarlane (2008) puts it, toilets are a symbolically charged issue in Indian slums.

Feminists, urbanists and queer theorists in recent years have drawn attention to toilets as spaces of representation, gender division and normative assumptions around sex/gender and embodiment. Overall (2007) makes the interesting point that in the Western world there is fairly widespread agreement that a person's sex is irrelevant to their rights and responsibilities in almost all contexts, yet the sex segregation of public toilets is largely taken for granted and situated in the 'design and management of the urban environment; of larger assumptions about sexuality, reproduction, and privacy that govern that environment; and of continuing compulsory sex identification and segregation which still define key areas of "public" space' (Overall 2007: 71–2).

What she suggests is that sex segregation of public toilets is a microcosm of the operation of sex and gender norms, which 'instantiates beliefs about danger, purity, privacy, heterosexism and the significance of biological sex differences' (ibid.: 75). It is a practice which is justified on the basis that it is what women and men want to meet their specific needs. But behind this lie assumptions about women's needs for privacy or bonding or caring for children, or gendered notions about what is acceptable in public, for example, where breastfeeding is concerned—that it is shameful or obscene to perform in public. Rather, Overall makes the case against sex-segregated toilets on the grounds that parents of either sex may have children of the other sex and that sex-segregated facilities create problems for gender and sex nonconformists.

These arguments are developed further in *Queering Bathrooms* (Cavanagh 2010: 4) which explores the cultural politics of gender and excretion, the moral panic and anxiety associated with transgender use of toilets and the ways people become territorial and defensive about the gendered composition of the toilet. In Cavanagh's (2010: 5) view, 'obsessive investment in 'urinary segregation'.. by gender is about perceived threat to sexual difference and to heteronormativity such that the 'the institution of gender-neutral toilet designs is an urgent and important political project to ensure access for all who depart from conventional sex/gender body politics. In the contemporary landscape, the toilet is a site where gender variance is linked to dirt, disease and public danger' (ibid.: 7). These links between sex and toilets are by no means new where men are concerned, with a long-established recognition, leading to heavy policing in many instances, that male toilets at times have constituted a site of gay men's cottaging (Houlbrook 2005). The argument that the architectural design of toilets plays a part in supporting the illusion that there are two binary genders which are visible, identifiable and natural, finds its parallel in the argument that the design of toilets also acts to exclude people who are frail, or disabled through restricted entrances or turnstiles (Molotch and Noren 2010).

This recent politics around public toilets takes another form in China. As Molotch (1988: 129) has argued the distribution of space in toilets produces unequal results due to women's different needs, such that only an asymmetric distribution of space would provide equality of opportunity. In China a re-versioning of the Occupy movement—which has typically occupied financial districts in global cities to expose wealth inequalities—has emerged to challenge the inadequate facilities for women. Unlike the 3:1 gender ratio recommended for public toilets in Taiwan, and 3:2 recommended in Hong Kong, across China as a whole a 1:1 ratio is the national standard for men to women's street toilets (La Franiere 2012). In 2012 after a long bus journey and encountering a long wait for the toilet as she watched men enter and exit the adjacent toilet with no delay, a local management student Li Tingting in Shanxi Province adopted unusual tactics to press for change. In the southern Chinese city of Guangzhou, she and half a dozen other activists hijacked the men's stalls at a busy public toilet near a park, warding off the men for 3-minute intervals every 10 minutes, and inviting women to use the vacated men's stalls. The operation, dubbed 'Occupy Men's Toilets', ended after an hour with, according to Ms. Li, greater public awareness. Since the following March, the ratio of men's stalls to women's in all new or renovated public restrooms in Guangzhou has been set at 1.5:1. A similar occupation in a public toilet at a long-distance bus terminal in Beijing was greeted by ten officers and three police vehicles.

The issue of public toilets is not new in China; the World Health Organization estimates that tens of millions of Chinese have no access to toilets and defecate in the open. A 2010 report estimated that 45% of Chinese lacked access to decent sanitation facilities. Nevertheless, China's sanitation has improved drastically in recent years with household acquisition of 19 million toilets a year and the hosting of World Toilet Organization's 11th World Toilet Summit and Expo on Hainan Island leading a national official to claim that China was in the midst of a 'toilet revolution'.

It is clear from this brief interrogation of the provision of public toilets and sewerage systems that they both illustrate and enact prevailing assumptions and notions of bodies and gender, as well as class and race. Implicated here are shifting norms of appropriate behaviours in public and private, which themselves help create those very boundaries. Public toilets also reflect and constitute specific socio-cultural and economic materialities of spaces at different times, where in some places basic corporeal needs are simply not met.

Laundry Practices

> The reproduction of 'appropriately' cleaned clothing is best understood in terms of a system of sociotechnical systems that co-evolve together. (Shove 2003)

Unimaginably large amounts of laundry produced in cities, from the clothes and sheets of private homes, to the tableware, towels and bed linen of hotels, have had an impact on city life, public space and sociality that has passed unnoticed. The fact that changing washing technologies and practices have rarely been constituted as a matter of concern by urbanists, despite their centrality to everyday life in the city, reflects the lack of importance paid to feminized domestic activity, a point consistently raised in feminist work on gendered divisions of labour over several decades (Beechey 1979; Oakley 1972; Barrett 1980). It is no coincidence that the idiom 'airing your dirty laundry in public' is deployed to describe revealing aspects of your private life that should remain secret.

As with the technologies and infrastructures associated with cleaning bodies and toilet practices, shifting laundry practices and technologies associated with this mundane object have over time summoned different spaces, socialities and socio-spatial assemblages in the city, enrolling different actors and multiple publics and constituting different associations, networks and relations in its wake as it travels from the home to the laundry and back again. The shifts and changes in washing practices, enabled by mechanization—itself a reflection of changing labour patterns and costs—have shaped and reshaped public/private boundaries in the city, as well as impacting on high streets and suburban areas where these activities have been concentrated.

'Private' Laundry: From the Home to the Streets and Back Again

Domestic clothes and linen washing practices in Europe and the US over the last century or more have taken a variety of forms articulating different gender/class/ethnic relations and private/public spaces as technical innovations in the industry changed. For women in wealthier households in the early twentieth century, and for some even later, dirty washing magically returned clean from hours of the hidden labour of domestic servants or washer women in private homes who earned around 3 shillings a week, with enhanced earning power of 3 shillings a week if they were in possession of a mangle (Mayhew 1861: 306). Laundry practices also took a more visible form; where a stream or river was close by, the women took the washing there gathering with others in a communal form of employment (Sidbury 1997) resembling contemporary practices in much of the Global South. Essential materials for washing at home included a tub of hot water, a washboard initially constructed of wood and later fabricated in metal, and a bar of laundry soap, or a dolly tub with a dolly stick (like a peg) to stir the washing, and a mangle or wringer. Limited supplies of soap meant economies of use, at least until the latter part of the nineteenth century (Old and Interesting 2013), and everyday linen might only be washed with ash lye, especially in poorer households and was typically performed by women. By the early twentieth century mass-produced tongs replaced sticks, and wet washing moved from public to more private, but still visible, sites to dry as clotheslines and pegs in back yards and gardens took the place of drying on trees, banks and bushes. Photographs and paintings of the growing industrial cities are littered with lines of washing fluttering in the wind.

The shift of laundry from private to public is said to have originated in the considerate actions of one Irish migrant Kitty Wilkinson in Liverpool, who, during the 1832 cholera outbreak, invited people in the neighbourhood with infected clothes to use her boiler, earning her the title 'Saint of the Slums'. Ten years later drawing on public funds her efforts resulted in the opening of a combined washhouse and public baths, the first in the UK. As discussed earlier, growing concerns about the sanitation, cleanliness and public hygiene of the urban poor led to the construction of municipal washhouses across the UK following the Public Baths and Wash Houses Act of 1846, many of which remained as an essential public service until the latter part of the twentieth century. Typically, for poorer households, dirty washing and its associated practices remained closer to home particularly in the public and philanthropic housing sectors, where laundries were constructed as an integral part of the estate. The first Peabody Estates, which opened in 1864, were built with communal facilities including shared sinks and WCs on landings, and bathhouses and laundry blocks with washing tubs and drying cupboards. There were three designs for the laundries—an outside block, a laundry across the whole of the top floor serving 22–23 flats, and partly open to the elements for drying purposes, or one on each floor containing tubs and drying cupboards for the flats there to share. Similar accounts are given as to the significance of these communal facilities for women's sociality. According to the archivist at Peabody, despite appreciating the self-contained facilities after the modernization of the blocks during the 1950s–1970s—many of the tenants described missing the contact with their neighbours.

By the turn of the twentieth century, many Londoners continued to live in crowded courts with no internal water supply and shared standpipes and outside lavatories were common. Even when water was piped to a house, there was often only one tap in a scullery, shared by all tenants (Museum of London 2013). Public baths and washhouses provided hot water and laundry facilities, where the washhouse supplied large tubs for washing clothes, as well as mangles and driers, and these became important sites of sociality for women as they carried out the family's laundry.[1] By the twentieth century power-driven washing machines began to

---

[1] http://www.museumoflondon.org.uk/Collections-Research/Research/Your-Research/X20L/Themes/1382/1202/.

replace the old washing tubs. According to one George Hargreaves who worked with Bradford and Tullis—the main suppliers of washing machines to local authority laundries—the public washhouses were, in effect, the original launderettes (Goodliffe and Temperley 2009: 89).

Changing technologies in the industry from the late 1900s had a profound impact on both domestic life and city spaces. Mechanization came late (arguably due to its gendering), shifting from a cottage industry to the power laundry between 1870 and 1914, as steam power and the commercial development of steam heated flat work machinery and mechanical rotary washers enabled large quantities of washing to be undertaken at the same time (Goodliffe and Temperley 2009: 5). This shift of domestic and local laundry practices, to the commercial laundry, the privilege of the higher-income classes, had distinct social and spatial effects on the city, providing new sites of sociality and conviviality for the laundry workers, and in the case of the US, new racialized labour relations.

The Rise of the Launderette

At the end of the Second World War the importation of the coin-operated washing machine from the US summoned new gendered socio-spatial relations and a new urban landscape into play. A prevailing emphasis on the nuclear family and pressure on women to create the perfect domestic suburban home (Wilson 1980) after six years of relative freedom from domestic drudgery during the war, created a fertile environment for the American company Bendix to import the coin-operated machine. The first launderette in the UK was launched in Queensway, London in 1946, and was an immediate success attracting 800 customers in the first five weeks (Bloom 1988: 14).

Bendix Company, who held the initial monopoly in the industry in the UK, controlled their expansion through the 1950s ensuring that each launderette was only one mile apart (Goodliffe and Temperley 2010: 89). By the mid-1950s launderettes had received widespread acceptance and 500 coin-operated launderettes, now also supplied by other manufacturers such as Westinghouse and Whirlpool, were to be found across the UK. The changing technology and ease of access to local launderettes was accompanied by shifting attitudes to washing. Addressing the 1958 Annual Conference of the Institute of British Launderers, the Director asserted: 'Not so long ago there was considerable pressure on the housewife to do the same as her neighbour; and to send all her household articles to the laundry. Certainly…she would not wish to hang her washing out on the line for all to see. But nowadays all that has changed and I cannot think of anywhere … where washing cannot be seen hanging out, and where the housewife is bothered in the least in seeing it hanging there, indeed one even sees it in the better class districts, and on Sundays as well!' (Ibid.: 90).

The rapid rise of the coin-operated laundry-the launderette mobilized a new set of socio- spatial and economic relations in towns and cities. Washing, hitherto a privatized activity consigned to the home (or close by) or commercial laundries, and invisible like much of women's work, takes on a public face, marking the high street with its presence. On every city street, at any time of the day, a group of mostly women could be found sitting by a washing machine, rubbing along in the same space in casual encounters (Watson 2006) or engaging which each other or the manager in animated conversation. At the same time, investment in launderettes provided a new form of small business investment—1500 were owned by single family units in Britain in 1968 and were particularly popular in industrial areas with three shift working hours (Mitchell 1963: 7). By 1975 a peak had been reached of 8400 units across the UK. Such were their success that the new industry engaged in continuous processes of refurbishment and modernization as illustrated in an Industry manual in 1963: 'Many came into being in the 50s—their design at the time seemed modern and up to date—just like the coffee bar. But just like the coffee bar of 1953 with its fake rubber plants, bamboo screens and Spanish bull—fighter posters, looks tatty and old fashioned in 1963 so some of the original self- service laundries with their simple damp—wash service, their tungsten light fittings and their utility décor now appear thoroughly "old-hat"' (Mitchell 1963: 56). From this writer's perspective diversification and innovation were far more common in the US, where launderers had introduced shoe repairs and even beauty parlours and coffee shops into the site—a far cry from the 'one man launderette business in a British high street, with its 12 year old machines, mouldering paintwork, fly blown posters and an elusive stench of old clothes' (ibid.: 60). A simultaneous development was the incorporation of laundries into public housing estates, often on the ground floor of council blocks for the use of residents, and frequently subjected to vandalization and limited maintenance.

My argument here is that not only did launderettes shift a gendered activity from the home to the street, which enabled the potential de-gendering of the practice as private chores became public, they also constituted a new form of public space in towns and cities. Launderettes notoriously were spaces of interaction, with shifting populations, atmospheres and intensities from day to night as students and single people, replaced the largely female or older populations of daylight hours. Though not typically recognized as such, these were quasi-public spaces of previously domestically performed work, which through the emergence of the coin-operated washing machine and tumble drier, and associated time needed for the task to be performed, assembled washers in casual relations of sociality and encounter. This social-material configuration also contributed to the shifting of social reproduction out of the 'home' into socially public, and economically private spaces of the city, assembling new familial, gendered, economic and social relations as domestic labour was reallocated to the private market, or to women's 'free hours', enabling the greater incorporation of women into the workforce in greater numbers.

So significant were these spaces of imagined possibility and encounter, often sexually inscribed, that they found their way into numerous instances of popular culture, from the song by the Detergents in 1963 *Leader of the Laundromat* (a parody on the Shangri Las song 'Leader of the Pack), and Coin Laundry—a song performed and written by Australian singer–songwriter Lisa Mitchell about finding love at the coin laundry; to the launderette in the sitcom East Enders which was a central focus of life in the community. The launderette didn't just feature in songs and soaps, in 1985 Levi's launched a now famous advertisement where a sexy young man exhibiting retro chic walks into a launderette to the lyrics of Marvin Gaye's 'Through the Grapevine' removes his Ray Ban sunglasses, casts an alluring gaze at the other customers, and seductively takes off his jeans and places them in the washing machine, apparently leading to a 20-fold increase in sales figures of 501 jeans in Britain. As Sir John Hegarty, the creative brain behind the ad, later described it:

'We wanted an egalitarian environment, somewhere you would find almost anyone, and the launderette had that' (Khan 2010). While the more cosy or parodic representations of everyday life in the launderette were given a further twist in the British film *My Beautiful Launderette*—a 1985 British comedy-drama film based on a screenplay by Hanif Kureishi, which depicts the reunion and eventual romance between Omar, a young Pakistani man living in London, and his old friend, a street punk named Johnny, tackling homosexuality and racism during the dark days of Thatcher's Britain.

By the mid-1980s, the growing affordability of washing machines and tumble driers signalled the gradual demise of the launderette as a commonplace feature of the British high street. According to the National Association of the Launderette Industry numbers in the UK peaked at 12,500 in the early 1980s dwindling by 2012 to 3000 across the UK. Unlike the earlier shift of washing from private to public or commercial space, this shift did not derive from technological change. Rather, it reflected the new prevalence of a mundane domestic object, the washing machine, in the domestic sphere, as purchase costs diminished, on the one hand, and repair costs for launderette washing machines increased on the other. The move of the machine into the home was also entangled with changing gender relations, as more women entered the workforce full time (militating against regular visits to the launderette), and new expectations of cleanliness meant at least two to three family washes per week. Pink (2007) suggested that domestic laundry practices also constituted a route to satisfy a 'quest to create a home and gendered self they [women] believe is morally satisfactory'. Cowan (1983) similarly saw domestic laundry as reflecting an enduring commitment to the preservation of practices regarded central to family life. The penetration of the home by washing machines was firmly in place by 2003 when Shove (2003: 117) found that the average British washing machines were used 274 times annually (392 cycles in the US) and washing machine ownership had reached 98% of all households (92% in the US). As Shove argues, domestic laundry practices are continually framed by typologies and classificatory frameworks creating new habits, as systems are held together through the coordination of materials and meanings by the

Photo 5.1 A council laundry—early 1900s. Sophie Watson

people who carry out the washing (140–1). The space of the public launderette, of shared machines and facilities, I suggest, affords lesser potential for such re-scripted practices (Photo 5.1).

Launderettes in public and social housing estates also went into severe decline over the latter decades of the twentieth century, as increasingly these spaces had become neglected and vandalized, leading to the installation of surveillance cameras, and the infrequent use of machines as tenants took their custom elsewhere or installed washing machines in their flats. This public shared space for low-income tenants now long gone has more recently been adapted for other uses. In Southwark for example, on the Kingswood Estate the council have adopted a strategy of converting the old laundries to create new homes. Councillor Ian Wingfield, cabinet member for housing, said, 'This is a brilliant, innovative scheme.… (which) literally creates space for homes from nothing. It's difficult to believe that what were such dingy, unused spaces have been transformed into such bright new flats, which will very soon be let to tenants.'[2] This demise of the high street launderette in the UK is nevertheless a spatially differentiated phenomenon, with launderettes still in evidence in medium/high-density areas dominated by low income or student housing.

Though these launderettes remain in some city spaces, they no longer represent a site of sociality and encounter, with the growing practice of service washes and bag drop-offs. Where customers stay they sit with laptops or magazines, while doing their wash, and the only form of sociality is between customers and laundry owners who engaged in familiar banter with regular users. John Trapp, owner of Associated Liver Launderettes in Liverpool, the UK's largest chain (Khan 2010), describes their contemporary more polarized customer base:

> We have people at both ends of the scale, from newly arrived immigrants with no access to hot water in their properties, to busy working couples who might have a machine at home, but just don't have time and prefer to have a service wash. Then there is the one thing that everyone owns that none of us can wash at home—a duvet. That brings most people to a launderette at least twice a year.

Despite their apparent demise, there are scattered attempts across the UK to revive launderettes as opportunities for social enterprise or a community hub. The Hilton Street Launderette in Manchester's northern quarter, for example, houses high-speed computers alongside washing machines, and provides coffee and sofas, to attract those who want to play games or watch films online while waiting for their load. While also in Manchester at the Clean Machine on Withington Road, during the summer of 2010 the launderette was transformed into an art gallery for a new exhibition by a local artist (Britton 2010: 2).

New York launderettes—laundromats in the local idiom—offer a distinct contrast, not least in their abundance due to high land values and the dominance of apartment housing where restricted space, money or regulations limit the prevalence of domestic washing machines. Laundry services in New York have largely been the domain of Chinese since their large-scale immigration to US cities in the nineteenth century with the

---

[2] http://www.southwark.gov.uk/news/article/161/from_filthy_laundries_to_fresh_new_hidden_homes.

5 Embodied Water Entanglements: Sex/Gender, Race/Ethnicity... 125

Photo 5.2 A wash and fold laundry in New York. Sophie Watson

majority of laundromats managed by Chinese families who perform bag wash, and customers express strong affect with respect to their quality with reams of posts on web sites where (sometimes racialized) comments like this are common: 'I picked up my laundry with trepidation …No weird stains! No holes, no grey whites! I was dumbfounded. I love these guys with the unreserved affection I have for smiling, friendly Chinese owned-family run businesses'.

Self-service laundromats in New York are organized around the concept of wash and fold, where large wooden boards for folding occupy the central space, at which customers stand in silence folding their washing while watching large screen televisions overhead (Photo 5.2). These are bustling places but where sociality is at a minimum, with no chairs or space for sitting down during the wash, a device in part to deter the homeless. Despite this, web posts suggest a high level of emotional investment in these local sites of domestic reproduction. A local NY journalist (Moore 2012) described her experience thus:

> I live right down the street from the laundromat but like everything in New York, going there means competing with everyone else for the washer. It means there are 25 washers in the joint but only 5 of them work at any one time…it means figuring out the timeframe when the number of people in there will be the lowest…it means not making eye contact with people as they are putting their dirty underwear into the wash…by the way how weird is it to fold your clothes in front of a group of strangers? You watch people fold their stuff secretly judging their character on the basis of their underwear.

Starkly reflected in these comments is the ambivalent affect associated with making intimate bodily matters public or visible. Commercial laundries are located out of sight on the fringes of cities. In summary, then, laundry practices of the household and the enactment of domestic tasks both shape and reproduce bodies on a daily basis and summon specific socio-spatial assemblages in the city.

Commercial Laundry—Dirty Washing Goes Public

I turn now to the commercial laundries, which represent the most public face of dirty washing and its transformation into clean objects. The advent of the steam-powered laundry in the mid-1850s had a profound effect on the urban landscape of the industrializing cities. Laundry collection, by horse-drawn carriages followed by motor-powered vans, became an increasingly visible part of everyday life in towns. As Bell (1900: 10) described the trade:

> Considering how the laundry trade has grown of late years by leaps and bounds, it would be a difficult matter to find a town, however small, worth of the name without a steam laundry, and the very first and most important outside consideration is a good horse and smart van. This should not be gaudy, but neat, for instance, a black or chocolate ground and gold letters, or a cream ground and crimson letters, or electric blue ground and deliver letters…. The chief point with regard to him (a smart man in livery) is a good character for sobriety and honesty.

Laundry buildings, containing large machinery for washing and drying, were striking features of the built environment, typically on the edge of cities, while over a dozen laundry machine manufacturers sprang up across the UK (Goodliffe and Temperley 2009: 4). Social shifts intersected with technological and material changes as the growing middle class in cities sent their washing to the power laundry. High levels of set up capital required local investment, but dividends were good, and local wealthy individuals saw them as a good speculative risk. As the prospectus for the Crouch Hill Sanitary Laundry Limited near Sherbourne pointed out:

> The profitable character of well-conducted Steam Laundries is well known, and careful enquiry into the returns of these undertakings shows that as the work extends the proportion of profit is increased. It should be borne in mind by intending investors that they will not only have the advantage of their washing being efficiently done, but also that the cost will be materially reduced by the handsome dividend anticipated upon the shares held in the Company. (ibid.: 10)

Over the following decades the number of commercial laundries increased across British towns and cities, located in suburban areas. Not only were cities visibly reshaped by the physical infrastructure and transportation practices resulting from this growth, so also new opportunities emerged for sociality in public space, not now in the washing houses or streams of the earlier period, but in the spaces of work associated with the trade as I discuss below. As a place of employment, laundry remained women's work being considered too demeaning for men, though with the growth of the power laundries a recalibration of gender relations emerged, as men took over the ownership and management of laundries (Mohun 1999) and involved themselves in the more specialized mechanical parts of the work. Driving the vans became an entirely male preserve with photographs from the 1930s showing men dressed in smart uniforms decorated with brass buttons standing proudly by their vans.

In the US, race added another dimension, where steam laundries across the cities and towns of America were operated by Chinese men from the nineteenth century, with a further gendered and racialized shift as changing technologies recast the industry as mechanical, scientific and manlike, and white male power laundry owners competed with the

Chinese steam laundry men to assert their authority and superiority (Wang 2002: 54).

The use of commercial laundries in London by middle- and higher-income households remained widespread through to the 1960s with laundry vans collecting or delivering laundry boxes a constant marker of wealth in the better off residential areas of cities, freeing housewives from this aspect of domestic drudgery. From the start of that decade their use by private households went into sharp decline precipitated by three factors. The first reflected the intersections of urban/rural life and class in unexpected ways. Typically, higher-income households in country areas delivered hampers containing the bed linen, towels and tablecloths to the local station to be dispatched to the laundries in towns and cities by train on a weekly basis. Thus, socially reproductive labour is made particularly invisible within middle-class settings became invisible as it was sent away and done anonymously, at the same time as offloading and occluding the consumption of a vital resource, namely water. In 1963 the Beeching report (Beeching 1963) aimed at restructuring the British railways, identified 2363 stations and 5000 miles of railway line for closure, representing 55% of all stations and 30% of route miles, with the stated objective of stemming the large financial losses incurred during a period of increasing competition from road transport. The reduction of the rail system had a considerable impact on laundries, which combined with the availability of cheaper domestic technology, and changing expectations around women worked to reduce their use. The family-run organization of the industry and their location on the edge of towns—in London the Ealing area was known as 'soap suds island'—represented further factors in their demise. As towns and cities expanded from the 1950s to 1970s the children or grandchildren of the original owners saw profits to be made in selling the sites for residential development—often now the sites of suburban housing estates and gated communities.

The use of commercial laundries has become extremely niche, essentially the preserve of A and B (upper and upper middle class) households living in the richer boroughs of central London—Mayfair and Central London, or country towns like Cheltenham, or catering to the hospitality industry. Blossom and Brown in Upton East London, which took over Sycamore, a company that has held the royal warrant for 200 years, is the most exclusive of the London laundries catering to the domestic sphere. Daniel whose family had owned the business over many generations described current practices and clients thus:

> [W]e became the only person doing private people in old laundry boxes and hampers—like the old ones. Going forward there will always be a niche—Mayfair, Kensington. My generation never knew what it is like to have your sheets laundered—whereas my parents all of them did this—sent their laundry off in black boxes with white writing—Sycamore—on it—very common then—people inherited linens—fabrics different in those days—good quality—it would last a life time—at the laundry it came back all nice and crisp—now rubbish quality wise—disposable items—throw them away—demand changes….People have dailies who iron for them—cheaper. … Old days we had gentlemen's handkerchiefs and socks—not coming through now.

Thirty per cent of their trade has remained in this sector, where washing (bed linen, table cloths and towels) is collected by their vans—still embossed with the old logo and dropped back a week later. At the same time new material forms assemble new washing practices. As Daniel pointed out, duvets have replaced the need for sheets, and can be made attractive as they are filled and have body, such that they cannot become easily creased. Duvet covers are also, he explained, not amenable to being washed in a commercial laundry, since the buttons and bordering militate against ironing or finishing through the large flat ironers. For Daniel the importance of high quality, well finished and packaged in hampers and cases—'how items are presented marks the distinction between good and bad laundries' and of diversification and innovation to keep the business viable was very clear. This company does the laundry for the House of Windsor (Photo 5.3).

With the demise of the domestic laundry sector, the proportion of laundry work for the service sector and industry has come to represent the majority of laundry work. Hotels, restaurants, hospitals, healthcare and other public services generate huge quantities of laundry, which is undertaken at a range of commercial laundries from large laundry groups such as Sunlight laundry to small enterprises across the country. Founded in 1900 in Fulham West London under the name Sunlight Laundry, like many other companies, it originally supplied domestic laundry services across the metropolis. It merged with another company in 1928,

Photo 5.3 Sycamore laundry boxes. Sophie Watson

expanding nationally, to change direction in 1963 with the rise of the domestic washing machine and the development of easy to iron fabrics, diversifying to launder and rent linen for the catering and hotel industries. Recalibrating the urban landscape once again, high urban land costs have forced this industry, where space is essential, to outer city areas. Sunlight headquarters are now located in a business park near Basingstoke. A similar trajectory has occurred for all the surviving laundries though often on a smaller scale.

In summary, rather than being an inert object of unpleasant matter, whose encounter with humans has been restricted to certain categories of person (poorer, female, or—in NY—Chinese) for its transformation to reuse, and thus passed unnoticed, dirty washing plays a vibrant role in making shifting socio-spatial relations in the city. What we have seen is that laundry practices have figured in producing and reproducing sex/gendered relations of labour, at home and away from the home—which have also been imbricated in distinctive relations of class and race and have had distinctive social and spatial effects. Doing the laundry has shaped and reshaped public/private boundaries shifting from privatized work in the home to the social spaces of the early washhouses, public laundries of the philanthropic and social housing estates, or later, of the launderette. As a private object made public through commercial laundry practices it became visible in the city in a different way, first in the commercial laundries scattered across the cities, and in its circulation in laundry vans on a daily basis, and later as a commonplace site in the launderettes of city high streets and in local neighbourhoods.

Conclusion

In conclusion, what I have suggested is that how and where bodies are washed, how and where bodies defecate, and how clothes are washed in the city is by far from a trivial affair. Rather, practices of cleanliness impact on the spatiality and sociality of cities. Yet the relative invisibility of these practices and lack of attention to their urban effects lie both in their gendered, ethnic and classed nature, and in the disgust or embarrassment we feel about dirty products that issue from, or are associated with, bodies (Douglas 1988). At the same time the socio-material and technological context of the city makes possible or constrains the possibilities for washing bodies and clothes. Historically in the West these simple tasks were difficult to conduct due to lack of running water and available facilities and appropriate sites, except for the wealthy. The impossibilities of keeping clean in many parts of the Global South remain. Clean bodies are thus constituted by and constitutive of city spaces in unequal relations of difference. These interconnected networks of bodies/water sites/buildings/technologies form an entangled web where cleanliness is enacted in different ways across time and space.

References

Allsop, R.O. 1894. *Public Baths and Wash-Houses*. London: E. & F. N. Spon and Chamberlain.

Barrett, Michele. 1980. *Women's Oppression Today Problems in Marxist Feminist Analysis*. London: Verso.

Beechey, Veronica. 1979. On Patriarchy. *Feminist Review* 3: 66–82.
Beeching, Robert. 1963. *The Reshaping of British Railways*. British Railways Board.
Bell, A. 1900. *The Steam Laundry with Which is Incorporated Laundry Wrinkles and Recipes*. London: Laundry Journal Office.
Bloom, Stanley. 1988. *The Launderette: A History*. London: Duckworth.
BMJ. 1911 Untitled. Anonymous author. 2. (2635).
Britton, P. 2010. Painter's Exhibition in His Local Launderette. *Manchester Evening News*, 2, June 1.
Brown, Goodwin. 1893. Public Baths. *Charities Review Archives Online* 2: 143.
Campbell, Agnes. 1918. *Report on Public Baths and Wash-Houses in the United Kingdom*. Carnegie United Kingdom Trust.
Cavanagh, Sheila. 2010. *Queering Bathrooms: Gender, Sexuality, and the Hygienic Imagination*. University of Toronto Press.
Cavanagh, Sue, and Vron Ware. 1990. *At Women's Convenience: Handbook on the Design of Women's Public Toilets*. London: Women's Design Service.
Cowan, Ruth. 1983. *More Work for Mother: The Ironies of Household Technology from the Open Hearth to the Microwave*. New York: Basic Books.
Douglas, Mary. 1988. *Purity and Danger: An Analysis of the Concepts of Pollution and Taboo*. London: Ark Paperbacks.
Economist. 2002. Bog Standards. *Economist*, August 15.
Gershenson, Olga, and Barbara Penner, eds. 2009. *Ladies and Gents: Public Toilets and Gender*. Philadelphia: Temple University Press.
Goodliffe, Brian, and Temperley Kitty. 2009. *A History of Our Industry: The Worshipful Company of Launderers*. London: The Worshipful Company of Launderers.
Goodliffe, Brian, and Temperley Kitty. 2010. *A History of Our Industry: The Worshipful Company of Launderers*. London: The Worshipful Company of Launderers.
Greed, Clara. 2009. The Role of the Toilet in Civic Life. In *Ladies and Gents: Public Toilets and Gender*, ed. Olga Gershenson and Barbara Penner, 35–47. Philadelphia: Temple University Press.
Houlbrook, Matt. 2005. *Queer London: Perils and Pleasures in the Sexual Metropolis, 1918–1957*. Chicago: University of Chicago Press.
Khan, Yasmeen. 2010. The Rise and Fall of the Launderette. *BBC News Magazine*, August 10.
Kristeva, Julia. 1982. *Powers of Horror*. New York: Columbia University Press.
La Franiere, Sharon. 2012. Chinese Women Demand more Public Toilets. *New York Times*, February 29.

Mayhew, Henry. 1861. *London Labour and the London Poor*. Vol. 3, republished 2008. Hertford: Wandsworth Publications.

McFarlane, Colin. 2008. Sanitation in Mumbai's Informal Settlements: State, "Slum", and Infrastructure. *Environment and Planning A* 40 (1): 88–107.

Mitchell, D. 1963. Starting a Self-Service Laundry. *Power Laundry and Cleaning News* (of which he is editor). London: Iliffe Technical Publications Ltd.

Mohun, Arwen. 1999. *Steam Laundries Gender Technology, and Work in the United States and Great Britain 1880–1940*. Baltimore: John Hopkins University press.

Molotch, Harvey. 1988. The Restroom and Equal Opportunity. *Sociological Forum* 3: 128–132.

Molotch, Harvey, and Laura Noren, eds. 2010. *Toilet: Public Restrooms and the Politics of Sharing*. New York University Press.

Moore, Madison. 2012. How to Do Laundry in the New York City. *Splice Today*.

Museum of London Archives. 2013. http://www.museumoflondon.org.uk/Collections-Research/Research/Your-Research/X20L/Themes/1382/1202/.

Oakley, Anne. 1972. *Sex, Gender and Society*. London: Temple Smith. Reprinted with a New Introduction. London: Gower, 1985.

Old and Interesting. 2013. *History of Domestic Paraphernalia*. http://www.oldandinteresting.com.

Overall, Christine. 2007. Public Toilets: Sex Segregation Revisited. *Ethics & the Environment* 12 (2): 71–91.

Penner, Barbara. 2001. A World of Unmentionable Suffering: Women's Public Conveniences in Victorian London. *Journal of Design History* 14: 35–51.

Pink, Sarah. 2007. The Sensory Home As a Site of Consumption: Everyday Laundry Practices and the Production of Gender. In *Gender and Consumption: Domestic Cultures and the Commercialisation of Everyday Life*, ed. Emma Casey and Linda Martens, 163–181. Aldershot: Ashgate.

Quekett, William. 1888. *My Sayings and Doings*. Kegan, Paul & Trench.

Shove, Elizabeth. 2003. *Comfort, Cleanliness and Convenience: The Social Organisation of Normality*. Oxford: Berg.

Sibley, David. 1995. *Geographies of Exclusion*. London: Sage.

Sidbury, James. 1997. *Ploughshares into Swords: Race, Rebellion, and Identity in Gabriel's Virginia, 1730–1810*. Cambridge: Cambridge University Press.

Skeggs, Beverley. 1997. *Formations of Class & Gender Becoming Respectable*. London: Sage.

Southwark Council. http://www.southwark.gov.uk/news/article/161/from_filthy_laundries_to_fresh_new_hidden_homes.

Taylor, Vanessa, and Frank Trentmann. 2011. Liquid Politics: Water and the Politics of Everyday Life in the Modern City. *Past Present* 211: 199–241.
VictorianLondon.org. http://www.victorianlondon.org/health/washouses.htm.
Wang, Joan. 2002. Gender, Race and Civilisation: The Competition between American Power Laundries and Chinese Steam Laundries 1870s–1920s. *American Studies International* XL (1): 52–73.
Watson, Sophie. 2006. *City Publics: The (Dis)enchantments of Urban Encounters*. London and New York: Routledge.
Wilson, Elizabeth. 1980. *Only Halfway to Paradise; Women in Post-War Britain 1945–1969*. London: Tavistock.

6

Public Waters: The Passions, Pleasures and Politics of Bathing in the City

Water is a substance that has a unique power to evoke passions, attachments and a sense of connection and belonging which enrols bodies in new socialities, alliances and politics in unpredictable ways. Water in cities is a site of considerable pleasure and delight enacted in a myriad of places from rivers to canals, from ponds to lakes and from harbours to beaches. Where water is present in cities it draws city inhabitants to it, to bathe and swim, or simply to gaze upon it in contemplation. Water is often regarded as having the magical power to cure ailing bodies or soothe sad and anxious minds. As Roger Deakin (2017: 4–5) puts it: 'When you swim you feel your body for what it mostly is—water… somehow or other it transmits its own self-regenerating powers to the swimmer. I can dive in with a long face and what feels like a terminal depression and come out a whistling idiot. There is a feeling of absolute freedom and wildness that comes with the sheer liberation of nakedness and weightlessness in natural water.' These powers of water engage people in fierce attachments to the water in which they swim. This chapter concerns the diversity of water spaces in cities and water's capacities as a vital material thing that generates particular passions, pleasures, attachments, and which brings people into new alliances and politics in unpredictable ways. In this I am building on a relatively new direction of research in public space, which has shifted the analysis of public space as a predominantly dematerialized realm of sociality, encounter and connection albeit in often liminal and marginal rather than formalized spaces (Watson 2006; Rhys-Taylor 2013), to the notion of public space as co-produced within networks of different actors, which may include humans, non-humans, objects and matter (Carter et al. 2011; Molotch 2010; Watson 2015).

In the debates on public space, until recently, the importance of water in constituting multiple publics has been remarkably absent from debates and investigation, with a few exceptions. A new direction of geographical research has articulated the significance of blue space, as opposed to green space. For Volker and Kistemann (2011: 449) *'the term "blue space" summarises all visible surface waters in space as an analogy to green space, not as a sub-category.'* Most writers to date have explored ocean and river waters rather than urban blue spaces; for example, Coleman and Kearns (2015) consider how blue space shapes the everyday life of living on an island in a New Zealand context which helps maintain the sense of well-being amongst older people. Foley and Kistemann (2015) explore swimming as a healthy body-water engagement in outdoor Irish swimming spots. In an urban setting Volker and Kistemann (2015) investigate the importance of urban blue spaces for health and well-being in two German cities. Another focus, particularly in Australia, has been the space of the beach (Game 1990; Booth 2001; Obrador-Pons 2007), where the beach is posited as a signifier of a national identity that rejects separations of class, as inherited from Britain, and gender but strikingly not race (Poynting 2006). In this respect a study of a beach in Darwin suggests that whiteness constitutes a force that exerts affective and wounding pressures on non-white bodies in hyper-visible public spaces constituting racially differentiated everyday experiences (Lobo 2014).

Here I draw more directly from Strang's (2004, 2005a, b, 2009) extensive and illuminating research which posits water as 'a vital "natural symbol" of sociality and of human-environmental interdependence' (Strang 2006: 155) and a site of belonging and attachment, something which she attributes to water's specific qualities of fluidity and transmutability which mobilize discourses and metaphors about flows and interconnections. Like Deakin, Strang points to the composition of human beings as 60%–75% water which promotes a particular affinity with this element.

Central to her research has been an exploration of different cultural groups' engagements with water in a diversity of contexts—the Stour river in Dorset (2004), the Brisbane and Mitchell rivers in Queensland (2009), and the ways in which these interactions mediate and constitute wider individual and collective identities within a complexity of social, cultural, economic, material, political and institutional relationships.

Comparing different water sites, Strang (2004) draws attention to cross cultural differences in cultural, spiritual, political and environmental meanings of, and attachments to, water while also suggesting there are universalities across time and space in the diverse interconnections between living organisms who are themselves composed of water in its myriad of forms. The major part of Strang's fascinating research into the identities and socialities constituted in human interactions with water have taken place in rural areas. Where her focus has shifted to the urban, she has explored water features in Brisbane as a material culture supporting practices directed towards establishing or maintaining community identities and celebrating social cohesion.

Much of the work discussed so far argues for the affective powers of water, connecting with a now fairly well-established set of literatures on emotional geographies. Some of this work engages also with insights from non-representational theory (Thrift 2007) with its attention to the experiences of things themselves and the active participation of non-human forces in events and the 'vital materiality' that runs through and across bodies both human and non-human (Bennett 2010). The entanglements of bodies with water make this approach all the more salient, since its very fluidity, vibrancy and transparency makes possible total immersion and oneness of the bodies that enter it. Anderson and Peters (2016: 4–5) similarly seek to demonstrate how the sea is 'alive with embodied human experiences' and a space which 'in and of itself that has material character shape and form… and is in a constant state of becoming'. In this vein Foley and Kistemann (2015), following Andrews et al. (2014) and their argument that non-representational theory can uncover, 'how the well comes into being', explores the experience of swimming as a well-being component of everyday life, where sea and the sky are deeply implicated in the production of feelings of well-being. Drawing on Bachelard's phenomenological notion of 'lived space' Game and Metcalfe

(2011) make a related argument in suggesting that it provides a relational alternative to Euclidean understandings of space as empty and inert, by foregrounding space that is both inside and outside, where emotions emerge not from the subject but from living space.

In this chapter I turn our gaze to watery spaces in cities, notably rivers, swimming pools, spas, lidos and rock pools, to illuminate the wider meanings of water imbricated in tastes and attitudes to health, bodies, cleanliness, class and gender distinctions, which as we see, vary across specific times and places. In so doing I draw on these vitalist/new materialist approaches, but also I put a significant emphasis on the cultural and historical, which are often overlooked, but which are equally important in understanding how water is enacted, seen, experienced, used and sited in the city.

Swimming

A love of swimming is not new. However, the form it takes, and the meanings and practices associated with swimming have changed over time, reflecting contemporary notions of gender, decency and bodies, inflected as these are by class and ethnicity. As far as we know, Everard Digby—a senior fellow at St John's College, Cambridge, was the first British writer on swimming. *In De Arte Natandi* (1587) he produced a collection of different drawings on all the possible strokes and techniques for entering the water, and states that his objective was to 'raise the art of swimming from the depths of ignorance and the dusts of oblivion' (Photo 6.1). For Digby 'this art of swimming, is also a thing necessary for every man to use, even in the pleasantest and securest time of his life, especially as the fittest thing to purge the skin from all external pollutions of uncleanness whatsoever, as sweat and such like, as also it helpeth to temperate the extreme heat of the body in the burning time of year' (Orme 1983: 117). But as Orme points out, there is a much older set of literatures from Roman times; Virgil, for example, extols the value of swimming to men for its capacity to harden young men to the frost and waters, while Caesar refers to both men and women bathing in the rivers (Ibid.: 603–4). Water bathing has been associated with a

Photo 6.1 In de Arte Natandi. Sophie Watson

diversity of benefits from hygiene and health, to relief from anxiety, or an escape from daily life. A swimmer at the Hampstead Heath Women's pond in North London told me: 'That moment when a kingfisher flies over my head stays with me later on the sweaty Northern line. I hesitate to say it's spiritual but there is something magical and peaceful about being here'. For Jessica Lee (2017), swimming the 52 lakes of Berlin through the four seasons represented a magical journey to heal her broken heart and depressed spirits, while, for others, swimming is neither a solitary activity, nor one of contemplation, but a space which forges connections with others, as we see shortly.

Rivers and Canals

During the twentieth century urban residents in many cities turned their back on their waterways as pollution and failing infrastructure turned flowing fresh waters into grimy, often infested and unappealing places to immerse one's body. Until this point there is plenty of evidence from literary and historical texts that people were swimming in the Thames whether it was the early Romans conducting military training swims or the Elizabethan noblemen. According to Davies (2012: 3) Edward II swam in the river in the eleventh century and Lord Bryon—well known also for his swim across the Hellespont—swam from Lambeth to London Bridge in 1807. Though the water was reputedly clean, the rapid currents and tides made the river a treacherous place, which led to the construction of floating baths at various points to offer protection. When Sir Joseph Bazalgette's sewer was completed in 1875, floating baths opened at Charing Cross. Swimming in the Thames was an unregulated affair until the mid-nineteenth century when a moral panic around indecency gave rise to regulation, and fines were issued to male bathers. As swimming gained in popularity, so also did the Thames become central to the recreational life of Londoners, and during the Victorian era clubs sprung up all along the Thames (Landreth 2017). Yet despite the success of young female champions such as Agnes Beckwith in the Victorian era, the clubs were male only spaces until the 1908 Olympic Games in London provided a renewed emphasis for women's swimming and the subsequent formation of women's swimming clubs. By 1957 the Thames had been declared biologically dead by the National History Museum and was considered too dangerous to swim in.

The last 50 years or so have witnessed a remarkable revival of rivers in cities across Europe and the US, with initiatives to cut pollution and curtail sewerage outflow turning the polluted waters of the tidal Thames into the home of more than 125 fish species. Symbolic of this transformation, and widely heralded in the media, has been the arrival of salmon—which were last seen running free in the Thames in 1833—and the much-loved British otter with the siting of otter life increasing over the last few decades. Once polluted tributaries, such as the river Wandle in the heart of London, are now so clean that they can support brown trout. This trend is replicated across the UK, where water conditions according to the Environment Agency have returned to those prior to the Industrial Revolution, and serious water pollution incidents have more than halved since 2001. Though such narratives are reproduced elsewhere, many cities in poorer parts of the world continue to face levels of pollution that require challenges and investment that can seem prohibitive, with pollution levels in the Ganges, for example, representing threats to human health and the wider environment.

The revival of urban waterways has opened up new spaces for river swimming and new swimming practices. The Thames engages London citizens in charity events like the actor and comedian David Walliams's 140-mile swim for Comic Relief, wild water swimming events, and individual adventures or publicity stints.

Nevertheless, the strong tides and eddies in the river remain a cause for concern leading to a new Port of London Authority by-law in 2012 which controls swimming in the busiest part of the Thames between Putney Bridge and Tower Bridge which requires prior consent from the harbour master. In Budapest where water has been central to urban cultural practices and different forms of sociality in its baths and spas, there are beaches on the Danube where improved water quality has also recently enabled swimming. In Paris the canal at La Villette is the first step in Paris's efforts to reopen its former murky waterways, and there is a plan to make the Seine swimmable again after 100 years in time for the Olympics in 2024. Restrictions on river swimming have always been contested. In Paris a small collective of urbanists (the Laboratory of Experimental Urban Swimming), for example, has organized late-night swims for several years in order to campaign for the reclamation of Paris's waterways. In 2018 New York is planning its first 30 bridges swim along the Hudson River. The river tides across the urban landscape are turning.

Swimming Pools

The Great Bath at the site of Mohenjo-Daro in modern-day Pakistan is thought to have been the first swimming pool, dug during the third millennium BC; the pool is 12 by 7 metres and was lined with bricks and covered with a tar-based sealant. Swimming pools were an important part of the early gymnasia in ancient Greece and Rome as part of the athletic training for young men in particular, and Roman emperors had their own private pools where fish were also kept—hence the Latin name for pools as 'piscina'. Highly decorated pools were also constructed in Sri Lanka in the fourth century BC. The construction of public swimming pools in England commenced after the introduction of the Baths and Washhouses Act in 1846, and pools were provided by local authorities in line with their own cultures and concerns; with greater enthusiasm amongst some municipalities than others; by 1918 Manchester was the largest provider of municipal pools in Britain. From their inception pools began to assemble a diversity of swimmers, formalized in 1869 by the establishment of the Amateur Swimming Association, and swimming clubs were attached to most swimming baths—for example in the London borough of Islington there were 100 clubs.

Writing in 1918 Campbell quotes a Doctor Stanley who maintained 'that for boys and girls "too much can hardly be said in favour of cold baths and swimming"' (Campbell 1918: 2). In his report, he extolls swimming as one of the best forms of recreation that is possible under cramped conditions which is widely recognized as 'a first-rate physical exercise owing to the muscular training it affords, and the tonic effect of contact with cold water'. In the year of 1914 at the start of the First World War, 4,445,729 swimming tickets were issued in England which represented an increase of one million tickets from the 3,277,160 that were sold in 1905. Swimming pools were another place where unequal gender relations were constituted, with reports of poorer facilities in the women's baths, or limited time slots allotted in the mixed baths. Women's bodies were constructed as troublesome for the management with the nap of their costumes or stray hairs said to contaminate the water, leading to outbreaks of gonorrhoea (Ibid.: 63).

By 1925 the separation of genders came under challenge at the reading of the Public Health Bill (20th March 181 cc2708-322708). Mr Womersley spoke of his fondness for mixed bathing:

> I am a swimmer myself, and so are my wife and daughter, and I do not want to have to go in one bath while they go into another. Fortunately, as

far as my own town is concerned, we have either ignored the Act or else got powers under a private Act, I do not know which, but we allow mixed bathing. I enjoy it very much, and I want other people to enjoy the same privilege, and it is only a reasonable thing that we should allow, at any rate, the municipalities to decide whether they shall allow mixed bathing or not without breaking the law if they do allow it.

Womersley's motion was seconded by a Mr Taylor, who—drawing on a sexist metaphor—nevertheless complained of 'the old ladies of the male sex' in his constituency who had convinced their local councillors that cold water was an extremely bad thing for people to indulge in, since it contributed to an 'outlook and a type of mind which it is not desirable to cultivate'. Throughout the twentieth century swimming has remained a popular urban pastime in Britain and many other countries where public and private investments make it possible.

Spas

Spas, thermal baths and Turkish baths have assembled urban citizens forging bonds and connections between a myriad of people, both local residents and visitors from afar travelling to take the waters for spiritual and physical healing. Spas and baths are a quintessential social space, which reflect a complex web of specific cultural meanings, local politics and histories. For example, the changes in Budapest's spa architecture reveal the connections between that region's Ottoman, Hapsburg and Communist pasts and local understandings of bodies and hygiene (Anderson and Tabb 2002: 4). For centuries spas in Europe attracted people in the belief that they cured disease or improved fertility, and that they had regenerative qualities for ailing minds and bodies; more often than not they were spaces of the wealthy (ibid: 3). As such, spas assembled an international elite of aristocrats, politicians and the bourgeoisie from across Europe, from Budapest to Buxton, Carlsbad to Bath. In recent years, they have found renewed popularity as mass tourism has fashioned water and its therapeutic qualities as a device to market hotels and resorts globally. Travel magazines, supplements and coffee table books are filled with images of the latest elegantly designed spas located in romantic settings.

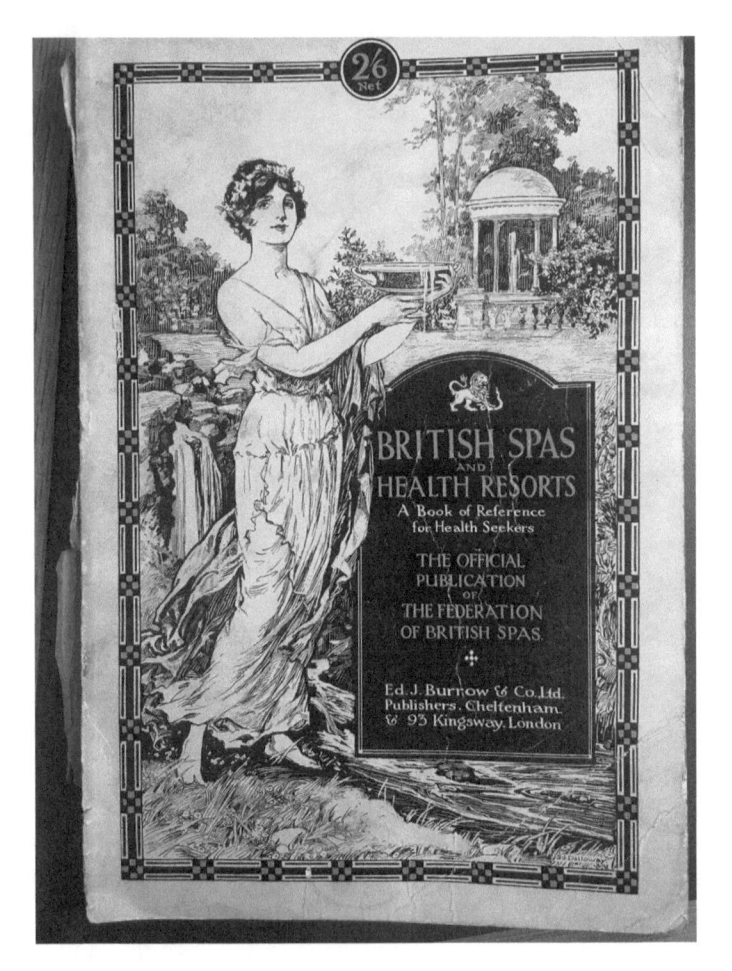

Photo 6.2 The report of the British Spas Federation. Sophie Watson

British spas offered medical cures drawing on the location's unique mineral waters and attractions. In 1916 a federation was established to support and market them; membership was open to local authorities; and the British Medical Association played an active role (Photo 6.2). At the time advances in medical science and new options for overseas travel increasingly had threatened their popularity as the English began to patronize spas across Europe. An early leading figure of the movement R.F. Fortescue bemoaned the exodus of British spa lovers to the Continental

Spas, which he attributed to the 'bloods of sea-rovers …within our veins' and a willingness 'to accept whatever comes to us from beyond our shores' and a little blindness to 'native worth' (Federation report page 2 undated). Rather, in his view, the British spas had the advantage of providing for active exercise due to the cold climate and the 'stimulant or sedative air' depending on the altitude or proximity to the sea.

For the next 100 years, the Federation of Spas had a colourful history, with the success of British spas rising and falling. First World War brought a particular interlude to spa culture, with the influx of between 50,000 and 75,000 war wounded and invalid soldiers for treatment at the spa resorts of Britain (Osborne 2005). Spas remained run down in the years post-First World War, due to lack of investment and a prevalence of an 'insular and chauvinistic approach' where inclusion of the overseas spas of the Commonwealth was seen by some to threaten the UK tourist industry (ibid). Throughout their history the medical benefits of spas were also a point of continuous contestation, with the Royal Commission on National Health Insurance before Second World War concluding that spa treatments were not to be recognized for statutory benefit despite petitions from the BSF. Notwithstanding their further exclusion from the National Health Service, advocates continued to advance their benefits with some success in the 1950s when the newly created Hospital Boards were allowed to contract for spa treatments with their local private spas. After 30 years of a chequered decline (with the exception of new facilities at Droitwich) revival came in a new form—the growing popularity of mineral water to drink and the 'health farm' offering leisure and health therapies.

By the turn of the century the fortunes of spas were entrenched with a new interest in treatment centres in the old spa towns, and growing health tourism where revitalized body cultures became a dominant theme in everyday life. Private investment in spas was on the increase, and two new industry journals were launched reflecting the growth of a multiplicity of body-conscious publics, mobilized in part through magazines, and more recently the internet. Traditional mineral spa therapies and cures were a thing of the past, replaced by a new tourist culture of leisure and relaxation, and alternative health therapies. A related site which assembled multiple publics was the Turkish bath, where the hot air in the bath is dry in contrast to the Islamic Hammam. In Britain the first such bath was opened in London in 1862, followed by a massive expansion of their presence across the country over the following decades. By the turn of the century there were more than 600 Turkish baths typically attached to the municipal public baths, and like many washing and swimming spaces at the time, segregated by gender (Shifrin 2016). Their popularity continued through the twentieth century, attracting a multiplicity of publics, including migrants from countries where the custom of bathing was more common. As local authority expenditure was cut back from the 1970s, Turkish baths gradually diminished, with only a few remaining, typically under the management of private companies.

Continental spas have had a less chequered history, where many of the spas which grew in popularity in the eighteenth and nineteenth centuries, such as in Germany and Austria, existed on sites where Roman settlers had established thermal baths earlier. Spas catered to a diversity of classes, and often provided one of the few available sources of medical care (Anderson and Tabb 2002: 25). In the second half of the eighteenth century the Hapsburg government promoted mineral waters for drinking and bathing as a route to improving general health, while the physician at Carlsbad spa introduced diet and exercise to complement the imbibing of, or immersion in the waters. 'Taking the waters' became a fashionable form of tourism, and new spa colonies in beautiful natural settings such as at Marienbad joined the newly refurbished older spas across Europe. Medical regimes provided the rationale for their existence, but equally important were the accompanying social life and daily rituals, drawing upper class patrons from across Europe facilitated by the railway construction of the 1830s and 1840s. In Hungary the baths in Buda successfully shifted their image as venues for lower class amusement, coinciding with the new Orientalism and a fascination with the titillating imagery of the Turkish bath and harem life produced in Western art works of the time (Switzer 2002: 157). The following century saw the construction of elegant baths in many European cities, of which many like the Szechenyi Baths in Budapest City Park, or the art nouveau Gellert Baths on the banks of the Danube remain popular today. Health tourism seems set to stay.

Outdoor Pleasures: Lidos

A special kind of swimming site is the Lido—a large public open-air swimming pool named after the Italian beach, which has been particularly popular in Britain and several other countries such as Switzerland and Hungary. These remarkable pools are very large—often 100 metres long, surrounded by expansive areas for sitting and basking in the sun, and other facilities such as saunas and cafes. The British Lidos display remarkable architectural forms, some like the Bristol Lido built in 1850 reflecting Egyptian designs influences, others were built in the 1930s when the construction of Lidos proliferated across urban landscapes and were thoroughly steeped in the art deco style designs. The Lido at New Brighton near Birkenhead in 1934 was said to be the largest pool in Europe with space for 2000 bathers and 10,000 spectators (Worpole 2000: 116). Many of the lidos fell into disrepair during the 1960s and 1970s as they suffered neglect from the local authority with closure affecting one after another in the latter decades of the twentieth century, as they were sold and redeveloped for other uses. During the 1980s alone eight of the seventy LCC in 1939 lidos remained. Campaigns in other cities have resisted their closure or fought their demolition with positive results. In Bristol for example, following the purchase of the Lido by the Sovereign Housing Association in 1998 for an undisclosed large sum (rumoured to be 81,500 GBP), and a long campaign, the Glass Boat Company was granted full planning permission in 2006 for the restoration of the pool and associated buildings leading to its reopening as a subscription pool with associated spa facilities, a poolside bar and a 75-seat restaurant in 2008—a rather different beast from its earlier incarnation.

Parliament Hill lido at the southern end of Hampstead Heath is illustrative of the vibrant public culture afforded by lidos. This lido (61 by 27 metres) was opened in 1938 as part of the London County Council's initiative from 1920 to 1939 to build 13 lidos to provide healthy recreation opportunity for Londoners. The Lido has separate changing rooms for men and women, each with an open shower area where most of the sociality is enacted. The London Residuary Body took over the Lido in

1986, followed by the Corporation of London in 1989, and was Grade II Listed in January 1999. An average of 50,000 visitors swims in the water each year. In 2005 the Lido was refurbished with a stainless-steel bottom, which sparkles in the sunlight. Like most London Lidos it is unheated.

This is a place that elicits passions. The, chair of the Parliament Hill Lido Users Group (PHLUG) eloquently described what her daily swim meant to her:

> When I'm not here, when I'm travelling with work, I miss it. You can't recreate it, … I think many of us found the lido at a time when we needed to find it. I certainly did and I know a lot of people who have come when relationships have broken up, when work has been bad. … And I always talk about the healing waters of the lido because I think they are in a way and I know … it sounds so hippy but it has absorbed so much negative energy from me over the years, I always leave feeling a million dollars. I might go in feeling like shit, I might have had a terrible time, I might be hung-over, but it will always make me feel better. So for that reason it is personal so when somebody says we want to do something to the lido or there's anything that they might want to change…, then I react because I want to save it, because …I know that it saved me and I know that it has saved a lot of other people as well. I think anybody that you speak to would give the same opinion, that it has a kind of almost mythical quality that just makes you feel better and has improved my quality of life enormously.

The intensity of this description vividly reveals strong connections between water and well-being, a theme that has been explored in different non-urban contexts also (Andrews et al. 2014; Coleman and Kearns 2015). This quote also reveals the power of water to encode meaning, and to be associated with the notion of healing, which is also often also gendered (Strang 2005a: 21–32). The Lido attracts swimmers across class, age and ethnic differences during the summer months when families come with children to spend all day at the side of the pool, mimicking a beach culture in the city.

Strong social networks are particularly striking amongst the early morning swimmers and who swim throughout the year between 7 am and 7.30 am. Gossip, banter and laughter resound in the early morning queue and in the waters and the showers after. Women shout from cubicle to cubicle, sharing news and information, teasing and joking with one another.

> As one swimmer described: 'We're always laughing. There was one time, we didn't mean to do it but we ended up doing a little experiment because Sue turned up without her swimsuit one day and she said: 'Oh crap'.
> And I said, 'Just borrow one'
> And she said, 'Oh I'm not putting on someone else's swimsuit and she got very upset and I said, 'Well what are you going to do?' And she said, 'I'm going to do an experiment. I'm not going to have a swim, I'm just going to have a shower and then hang out with you guys and get changed and I'm going to see what the effect is.'
> 'And actually she said that the effect was the same, with or without the swim. It was just the social collective thing, having a laugh.'

The fluidity of water connects bodies through swimming together every day, chatting as they go, laughing in the showers naked, and forges strong bonds, which carry over into mutual care when someone is sick, sending get well cards, or writing travel blogs from holiday. When a fellow swimmer is in trouble, news travels fast and support is provided. As Mary explained:

> I think it's also because when you swim at the lido, particularly in the winter, you look out for one another. There's that kind of … It's an unspoken rule but you're … You are on the lookout for checking that people are OK. So when old Sarah, Doctor Sarah we're talking about, when she used to swim or when Isabel, you know who arrives in the wheelchair, when they're swimming you just keep a weather eye on them just to make sure that they're OK and you know that there are people there that, if any of us got into trouble, there would be someone there to look after you.

Male swimmers reported similar levels of jocularity and connection. Richard, who swims at the Lido in the winter and the men's pond in the summer, put it this way:

> At the lido there's a lot of banter which goes on and a complete disrespect for age, character, etc., whereas at the ponds ….it's much quieter, even though it's

a bigger space and in some ways it's the enclosure where we get changed is more open, it's much quieter and you talk within your groups, it's not often that you'll call across to somebody else who is 10 metres away from you on the other side of the enclosure. So if there's 2 or 3 of you together then you might be talking but unlike at the lido you would call across to different cubicles and whatever and answer this voice that's called out and shouted something at you.

Such are the social connections amongst the morning swimmers, that Peter, a 64-year-old local (others come from as far as Croydon to swim each morning), organizes what he jokingly referred to as an annual 'works outing' during the summer, restricted to the year round swimmers. Twenty or so men and women set off by train to visit a lido in another part of the country and share a pub lunch. Talk of the event fills the air during the summer months.

The power of attachment mobilizes resistance as revealed in the response to a recent intervention by the City of London Corporation in 2015 when talks were initiated with the sports and leisure company Fusion Lifestyle to consider a possible takeover. Fusion's reputation was poor following their management of another Lido in an adjoining borough (Park Road in Haringey), which, according to a local swimmer had been shambolic (Banks 2015). The concern was that the pool would be closed during the winter months when fewer users render the pool less profitable. In the context of a drive to find new ways to raise income as the Corporation of London faces a 10% cut to its Open Spaces budget, this was considered as one option. This proposed privatization of the Lido, according to the Chair of PHLUG, enlivened the moribund users' group to action. Having been ignored by many other swimming campaigners on Hampstead Heath, in the spring of 2016, the recognition that the Lido was a precious resource that needed protection mobilized swimmers to vote in a stronger spokesperson to represent the Lido at the Corporation's Swim Forum. Since the renaissance of the group, there has been a vibrant politics of meetings and social events, drawing on Twitter, Facebook, Instagram and a new website, and the production of shirts and sweatshirts with the Lido emblazoned on the front to raise funds and the profile of the Lido as a significant public space. By 2017, 2000 GBP had been raised. The Lido has also been registered as a community asset, which means no decision can be made without proper consultation. New visions for the Lido now include redeploying the surrounding buildings for a yoga centre, a sauna and a gym to join the café already in place.

Outdoor Pleasures: Ponds

So far we have seen how water in place invokes passions, attachments, encounters and connections thus shifting debates where public space is predominantly de-materialized (Sennett 2010; Watson 2006). Urbanists have only recently been attentive to the liveliness of objects and matter in co-constituting public space, and the people who inhabit it. In what follows I consider how water in the ponds on Hampstead Heath not far from the Lido, and like many of the urban water sites so far, has constituted an active politics and produced lively networks, assembling not only those directly involved with water in an embodied sense, but also those who simply love the Heath and its water sites, as walkers and local residents. The first moment in our story of water's capacity to assemble new political connections and publics was the Corporation of London's attempt to restrict winter swimming in 2004. This was articulated as necessary to limit the risks associated with cold-water swimming. Simultaneously the introduction of entrance fees was proposed as a revenue-generating device. Both strategies met intense opposition as the long-established and well-organized swimmers drew on their professional expertise to resist the initiative, foreshadowing the even greater resistance to the construction of the dam more than a decade later. Unlike struggles to maintain municipal baths in low-income areas, these events brought together groups of well-educated and predominantly middle-class locals who had the time, resources and skills to act as powerful advocates and mobilize strong resistance.

Not many cities have ponds where swimming is possible. The ponds on Hampstead Heath represent another place which provides a colourful window onto the power of water to engender passion. These sites are also illustrative of a significant trend in the management of cities to deploy discourses of risk that have the potential to undermine the viability of public practices and public space. In 1889, the recently formed London

County Council took over the Heath, leading to fears that it would be turned from its uncultivated meadow-like form into a municipal park. The formation of the Hampstead Heath Protection Society quickly followed in 1897 to preserve the 'natural and wild' state of the heath for public to enjoy. In 1972 the London County Council was replaced by the Greater London Council (GLC), resulting in a renewed concern that the heathland would be tamed and the reinvigoration of the ailing Hampstead Protection Society (shortly to be renamed as Heath and Hampstead Society which is still in existence). With the abolition of the Greater London Council in 1986 Hampstead Heath was transferred to the London Residuary Body and, in 1989, to the Corporation of London (CoL) in whose hands it remains.

Water is crucial to the identity and embodied experience of this public space. There is a chain of fresh-water ponds, fed on the Eastern side, by the river Fleet, three of which have been used by swimmers since the 1880s. One is mixed and is the least private and most poorly located of the ponds—reversing the more usual trend of heterosexual privilege. The other two ponds are exclusively accessible to one gender and are hidden behind trees and surrounded by grassy banks. Each of the ponds includes wooden changing rooms and showers. There are many other ponds, some inaccessible, others designated for dog swimming, model boats or used by anglers—there are some 342 fishing permit holders on Hampstead Heath. These two ponds, particularly the women's pond have a rich history of attachment and affection, for those women and men who swim every day of the year, and for others who express delight at the beginning of the summer season, seeing familiar faces from previous years, or new faces who have been drawn to the ponds by word of mouth, or even TripAdvisor and tourist sites. As Esther Freud (2013: 33) puts it: '[T]here is a special bond between the swimmers—they smile at each other as they glide by and introduce their daughters as they come of age' (Photo 6.3).

The Kenwood Ladies Ponds's Association was established over a century ago and has been a force ever since. Many stories, poems and memoirs attest to the passions and pleasures the pond affords to the huge numbers of women who swim there. The current Chair of the Kenwood Ladies Association described a sense of how water creates a space for some kind of memory to form, either of a fantasy or a real moment, which

Photo 6.3 Hampstead Women's Pond. Katrina Silver

remains with the person as they travel through their day, giving them a feeling of calm and relaxation which survives the embodied and physical experience of immersion in the water:

> I just kept a diary every day of what was happening and it was just brilliant. And it's all that thing about how you relate to water, what sort of … I mean I remember one day coming back and thinking about being a mermaid. What is it that, the relationship between you as a human and the water and this myth of a mermaid.

These comments underline the idea of water as a space of entanglement where water enacts new forms of being human, and where the figure of the mermaid stands in for water/human hybridity. In my research on the ponds, what emerged was the power of water as a space of redemption and immersion away from the messiness and tensions of the everyday, and as a space of connection with the sublime and with tranquillity. This relates to the specificity of embodiment in water, where water and skin collide, with few barriers or protection to disrupt the assemblage. The particular presence of the water in this gendered space constitutes the surrounding grassy banks as a place where women can live their naked bodies (no tops are required) and display mutual affection openly without being subject to the male gaze. Women described their appreciation at swimming in the fresh water of the pond fed by a river and where chemicals are absent, which made them feel more connected to 'nature' than swimming in man-made pools. The all-year swimmers, especially those in the early morning—like the swimmers at the Lido—were particularly vociferous in articulating their passion for the ponds. But this passion is also prevalent amongst people who confine their swimming to the summer months, who represent a multiplicity of publics where women, and men at the men's pond, of all ages, ethnicities and sexualities congregate on the banks on hot days. At the ponds the water itself as fluid and unbounded matter enables easy chat, relaxation, laughter and teasing as women stand on the edge hesitating before throwing themselves into the cold water. This is more than a public space of mutual co-existence; rather, water constitutes the very sociality of bodies that are present. For women from minority communities, such as Hasidic and Muslim, where public display before men who are not their husbands is forbidden, the pond represents a rare opportunity to swim in an open-air space.

This sociality at the ponds is legendary (Griswold 1998) and takes different forms across time and seasons. Many men and women have swum there all their lives, and many are local actors, lawyers and others—part of a dying community of Bohemian residents, who are gradually being displaced as property prices escalate through overseas and city investment (Webber and Burrows 2015). Some of the time, particularly during the summer, the men's pond is a popular meeting place for gay men from across London. According to one swimmer at the pond: 'It changes as the day goes on, so when I go there now it will be predominantly gay but certainly in the morning there's one or two but they're there for the swim, they're not there to display themselves or flirt or court or whatever, they're there because that's what they do.'

When the pond was under threat from the construction of the dam on the Heath, a group of women assembled to knit a scarf long enough to encircle the pond, which subsequently was transported to span the meridian at Greenwich—reflecting an alternative—and often feminized—spirituality. Others who are involved in the women's pond choir participated at a Water Aid event on the South bank of the Thames in 2016. On New Year's Day swimmers flock to the pond for a celebratory swim and lunch.

Water here also has the capacity to engender contestation as witnessed in recent debates on the inclusion of transgender women. In the summer of 2018 a vociferous group of women raised their objection to the inclusion of transgender women on the grounds that they changed the atmosphere of the space and threatened the comfort of women with strong religious beliefs. The Corporation was quick to respond arranging transgender training for the staff and endorsing the law which clearly states that transgender people can access single sex spaces that match their gender. As Stonewall, the UK's largest LGBT organization pointed out, much of the reporting of the issue in the media had been toxic (Weaver 2018: 7). The capacity of the ponds to engender attachment was similarly revealed when the Corporation of London in 2004 attempted to restrict winter swimming in the ponds on the grounds that cold-water swimming represented a risk, drawing on 'expert' reports which suggested that swimmers were liable to suffer heart attacks. Opposition from the winter swimmers was fierce, with swimmers asserting that the probability of danger was limited and arguing for freedom of choice to risk their lives in the ponds.

Engineering interventions by the Corporation of London in 2015 to construct a dam at the boating pond similarly revealed the vital capacity of water to constitute a multiplicity of political publics, unfolding new water-human interconnections in Hampstead Heath. Discourses of risk were central here (Beck 1992) as risk analysis came to dominate decision-making. The key legislation affecting the ponds and reservoirs are the Reservoirs Act 1975 and the Floods and Water Management Act 2010. Under the Reservoir Act the designation of a body of water over 25,000 cubic metres defines it as a reservoir. This initially only applied to three of the ponds on the Heath. However, the Floods and Water Management Act 2010 reduced the definition of statutory reservoirs to 10,000 cubic metres, and those in a chain with a combined volume greater than 10,000 cubic metres, effectively therefore affecting all the ponds on the Heath. A report was commissioned by the City of London from hydrologists, Haycock Associates in 2011 to determine the ponds' compliance with the two Acts, which concluded that during 'extreme rainfall events', the dams retaining the ponds on Hampstead Heath could not be relied upon to store the additional volume of water, thus potentially causing any excess to 'over top' possibly leading to a breach (COL 2013). If this water combined with the floodwater, there would be potential risk to life and property downstream. The large number of visitors to the Heath, it was argued, exacerbated this risk in summer, with the ground's consequent compaction and inability to absorb water. The report's conclusion was that the Probable Maximum Flood (PMF)—as it was called—under the conditions produced by a very large storm could thus lead to a catastrophe causing loss of life and damage to property downstream. The probability of such an event was recorded as 1 in 400,000, and like many engineering models was operating with uncertainties which 'represent an abstract and idealized version of the mathematical properties of a target' (Murphy et al. 2011). A subsequent detailed study was commissioned from another engineering company—Atkins, using computer-modelled results to assess the largest PMF that the dams could face and their ability to withstand it (COL 2013). Though this report estimated flood peaks at 30%–50% lower than Haycock it similarly concluded that there were potential risks to life and property downstream and that repairs were needed to ensure the safety level. Atkins then considered the different feasible engineering options on each pond chain alongside the environmental mitigation and compensation.

The precision of the figure denoting risk (if impossible to define accurately) became the red rag to a bull for those in opposition who emphasized the absurdity of the need for intervention. In her discussion of flood risk, Whatmore (2013: 39) refers to those 'moments of ontological disturbance in which the things on which we rely as unexamined parts of the material fabric of our everyday lives become molten. Such situations, matters or forces render expert knowledge claims, and the technologies through which these become hardwired into the working practices of commerce and government, the subject of intense political interrogation.'

These ponds for many years which had represented the 'material fabric' of the heath, where people swam, fished, sat in contemplation, were suddenly rendered a space of danger to be addressed. In this sense the notion of risk represented the rationale for the intervention and the political opposition that ensued. Once the figure entered the public sphere it became a key signifier and actor—a non-human force—in the events that followed, assembling voices of dissent. The figure, like many other similar predictions was open to contestation, since effectively it could not be known prior to the catastrophic event's occurrence. As Ritvo (2009: 177) argued in her history of Thirlmere reservoir in Cumbria: 'All modern environmental arguments rest on predictions—usually (although not always) about benefit to some people or about harm to landscape, flora, fauna, and other people.' Like all predictions, they are, by definition, unprovable—at least at the time.

From the Corporation of London's perspective, they were responsible for the reservoirs and as a 'risk averse organization' were required to act. What is interesting here, as in other public spaces, is that the articulation of a risk identifying a danger which may have been present for many years and invisible makes action inevitable, since however unlikely its occurrence, responsibility for failure to act would be clearly attributable. Even though the risk was low, the consequences were considered to be immense. According to the superintendent: 'Many of the properties downstream have basements. If the dam collapsed and if there were a cascade effect—it would be catastrophic—the predicted loss of life is 1000 people. It might happen once in 400,000 years—but if it did happen it would be terrible.'

Like many of the protagonists in favour of the intervention, the superintendent exemplified the precautionary principle which

> in essence enables, encourages, or requires policy makers to "err on the side of caution" by adopting relatively stringent regulations—even if the available scientific evidence of the risks posed by a particular business practice or product to public health, safety or environmental quality are unclear, inconclusive, ambiguous or uncertain... it enables policy makers to impose regulations on the basis of a potential or reasonable likelihood of harm, especially when there is a

possibility that the harms stemming from a failure to regulate may prove serious or irreversible. (Vogel 2012)

From the dam's inception, water's capacity to enrol fierce attachments and passion was evident in the strength of opposition from protesters who mobilized professional connections and networks, and press coverage. Several groups and blogs were established to fight the proposals including members of the Hampstead and Highgate society and 'Damn Nonsense', and national campaigns like 38 degrees joined the protest. Multiple publics, with unlikely alliances, were constituted through the process, including walkers, swimmers and fishermen, and interested local residents, inflected with the particular socio-cultural mix of Hampstead—artists, intellectuals, lawyers, 'bohemians', and the more recent influx of rich new home owners employed in the media, finance and multinational company sectors. Unlike open spaces in the constituency of South London that the superintendent had encountered in his former job as the director, where he described receiving only four letters in the course of a river restoration project on Surrey Commons, this was a vocal, passionate and confident public.

The objections were of several kinds. First, there was a concern that the concentration of power to affect the decision lay in the hands of one civil engineer—who was considered to be already partisan in that he was operating under contract from the City and was acting according to a document—'Floods and Reservoir Safety'—published by the Institution of Civil Engineers to which he belonged. As articulated in the Camden New Journal: 'The engineers who implement the "guidance" in Floods and Reservoir Safety and profit from works arising out of it, are the same engineers and their employers who drafted the guidance.' Many perceived this as corruption. Their second point was that the dams in their current form were very unlikely to fail; here the prediction that the storm that would cause the dams to flood might only occur once in 400,000 years was repeatedly mobilized. The third set of concerns was that the works would destroy the natural beauty of the heath with the insertion of material artefacts such as large concrete dam walls jutting into the landscape, which, despite their replacement in the final designs by less ugly material, remained fixed in the minds of the public and a point of contention.

6 Public Waters: The Passions, Pleasures and Politics of Bathing... 159

Three packed public meetings held in Hampstead, in January 2012, in Belsize Park in November 2013, and in Highgate in February 2014, overwhelmingly condemned the proposals. By the end of that year 13,000 had signed a petition against the works, and 905 people had written with detailed objections to the Planning Application. The local papers—the *Ham & High*, the *Camden New Journal* and the *Village Voice*—had published over 70 articles and letters against the project with headlines like 'Beautiful Hampstead Heath Is About to Be Mutilated to Satisfy Corporate Greed' (Marcus 2015), 'Hampstead Heath Ponds Project Is a Dam "Fiasco"' (Banks 2015) and 'Warning Shot Sent to City of London over Hampstead Heath Ponds Project' (Marshall 2014). What we see here in this eruption of widespread opposition is 'how environmental disturbances, like flooding or earthquakes, might 'force thought' among the people affected by them and, thereby, occasion new political associations and opportunities' (Whatmore 2013: 34).

Despite all the opposition, on 28 November 2014 the case was adjudicated at the High Court where the judgment ruled in favour of the works, and the Hampstead and Highgate Society were refused permission to Appeal. The final episode took place on the 15 January 2015 at Camden Town Hall when Camden Planning Committee made its decision to go ahead with the project, based on the AECOM report, in the face of almost unanimous opposition from the multiple publics present in the committee room. As one local campaigner wrote: '[N]ow the public has been asked by Camden whether they are prepared to sacrifice this treasured historic landscape to these extreme measures for the sake of a risk that may never happen; their answer has been a resounding "no".' But they are ignored, dealing a lethal blow not just to the Heath landscape but to democracy itself (Marcus 2015).

Once the work began, the local newspapers—particularly the *Village Voice*—referred to the devastation being reeked on the heath's natural beauty, a point consistently recorded in my field notes from conversations or overheard comments during the period, with little apparent understanding that nature is always the product of earlier interventions—'an artificial world, even in most areas of the countryside' (Mukerji 1997: 36). The information officer (CoL) articulated the irony of their position very clearly: 'Our reservoirs look like ponds, but they are reservoirs. They were built as such. But most people think of them as ponds. As natural. But the whole heath is managed to make it look natural—a huge amount of work goes into creating that effect—if it was not managed it would be scrubby and brambly woodlands.'

Nevertheless, as the project unfolded the opposition diminished for a number of reasons. Many had come to be persuaded of the need for the works to be undertaken, either through reading the reports or attending stakeholders meetings; others were resigned to the fact that the battle had been lost, while others were appreciative of what they saw as considerate construction protocols of the engineering company on site and the 'soft engineering' practices they had been required to adopt, or had had their negative preconceptions overturned once work was in progress. As one member of the opposition explained:

> A lot of people have assumed the worst—imagining loads of concrete everywhere…they haven't taken time to look at the plans some of them want to see the worst—a lot of them are pleasantly surprised.

Similarly there had been success in many of the demands made, such as the request that the reconstruction of the old changing rooms and swimming area was done in accordance with the Kenwood Ladies Pond Association 's wishes with the result that

> [i]t now looks fabulous… gorgeous. The water quality's great, the buildings are nice … when you go into the changing room bit and you walk through into the shower, the showers are at the far end. It's a beautiful big room, a lovely big space, big window there.

By 2017 the dam was completed different versions of the ponds had been assimilated into new practices and patterns of use, and the works were gradually forgotten as swimmers became accustomed to the new facilities, and dog walkers circled the boating pond along a different path. New layers of narrative, history and materialities had been sedimented. What had been revealed was the power of water to assemble people in their resistance to the dam project. As a result the parameters of the project were revised towards a 'softer' design, careful construction practices and new facilities such as the new changing rooms.

Outdoor Pleasures: Rock and Sea Pools

In several coastal cities pools have been hewn out of the rock, or constructed within the sea itself. These are often magic places, where water assembles bodies in particularly sensuous and stimulating ways, where salt, froth, waves, ripples, currents and untamed waters lift the urban citizen into a renewed sense of vitality or calm. The history of Sydney's rock pools is embedded in its colonial history of Australia's early settlement and the destruction of Aboriginal people and their land, when convicts were put to work transforming rocky outcrops into the pools that remain today. Others were built in the depression by unemployed labourers or made by wealthy Victorians in an era when it was considered immoral to swim during the day at the beach. Each of the pools has a distinct character and ethos, customs and practices that draw different sub cultures to them. Many of the pools have a gendered history and racialized history, constructed as some were on Aboriginal land.

A women-only bathing pool—McIvers—was built at Coogee on the South side of Sydney in 1886 by the McIver family who ran the baths until 1922 when the Randwick Ladies Amateur Swimming Club was formed. According to legend a pool here served as a swimming site for Indigenous women long before the 1880s. McIvers, like the Hampstead women's pond, is a key site for a diversity of women—particularly from communities where women are unable to display their bodies in the presence of men, or who prefer to swim in gender-segregated areas away from the dominant male culture of everyday life. Such claims by women have the capacity to enrage less enlightened men. In 1995 a local Coogee resident—one Leon Wolk—who resented the presence of a women-only pool disputed his exclusion and took his complaint to the Anti-Discrimination Board. Although McIver's was granted an exemption from the Anti-Discrimination act, this is due to expire in 2020 (Iveson 2003).

In the early years the ocean pools were gender segregated with separate hours designated for men and women, and bathers were required to wear costumes. Gender exclusions were inscribed in the social relations of the pools, with swimming carnivals and competitions restricted to men. The swimming club at Bondi—the Bondi Icebergs—was a classically

Australian male only preserve formed in 1929 by a band of local lifesavers who desired to maintain their fitness during the winter months. Included in the constitution was a rule that to maintain membership it was mandatory for swimmers to compete on three Sundays out of four for a period of five years. It wasn't until 1995 that women managed to successfully challenge their barring (Photo 6.4).

Just as the ocean pools have their distinct identities and cultures, so also do the beaches, drawing in different people enacting diverse practices. Bondi beach is represented as the quintessential Australian beach, attracting visitors from across the world, keen surfers and multiple publics from across the Sydney region. Along the coast from Bondi, Tamarama beach is seen as the cool and often gay male beach and known locally as 'Glamarama'. From there a mile or so down the cliff path lies Bronte a beach which is popular with families and small children. The harbour beaches similarly have a distinct flavour; Watson's bay, for example, is

Photo 6.4 The Bondi Icebergs Pool. Sophie Watson

known for its famous fish restaurant Doyles, and Neilsen park as a space where Italians congregate for picnics during the weekends in large family groups. There is no one substance that is more central to Sydney's identity than its water that connects and differentiates its citizens in a multiplicity of cultural practices and embodied pleasures. Sea pools are similarly constitutive of everyday life during Copenhagen's summer months, reflecting also the exquisite design for which Denmark is known. For example, the Kastrup Sea Bath, or The Snail as the locals call it, was built in 2005 entirely out of azobe, an African wood that is harder than steel and never rots or attracts woodworm and is illuminated at night by coloured lights.

Conclusion

In this chapter we have seen how the substance of water has a unique power to evoke passions, attachments and a sense of connection and belonging which enrols bodies in new socialities, alliances and politics in unpredictable ways, but nevertheless in ways which are embedded in prior histories and cultures. Water in its very substance appears as soft and sublime, as redemptive and spiritual, as connecting and enabling, as wild and cleansing, and as having the capacity to enhance a sense of well-being in those that swim in it. Immersion in water appears to make people feel at one with the world and care free, which I suggest derives from its very fluidity 'boundary-less-ness', vibrancy and transparency. These attributes mobilize a very particular sense of embodiment that gives public spaces a distinctiveness that engenders passionate attachments. Where interventions are made which threaten the inclusion of multiple publics in water spaces, water becomes an actor in enlisting active participation in the spaces and events that unfold. The vibrant matter of water runs through, shapes and flows across human and non-human bodies, where the agency that emerges, as Bennett (2010) suggests is the effect of unplanned and random configurations of human and non-human forces.

In summary, water matters in public space as a vital actor without which a diversity of democratic and diverse socialities would not take place. The very substance of water has the capacity to evoke a sense of attachment and belonging, which generate new connections and politics.

These connections and politics are also deeply cultural and social as we have seen here. As such, city governments and local municipalities need to value and defend water sites that exist in the locality, resisting their take over by private interests as spaces for the generation of profit, or their curtailment or closure through the dominance of discourses of risk.

References

Anderson, Jon, and Kimberley Peters. 2016. *Water Worlds: Human Geographies of the Ocean*. London: Routledge.
Anderson, Susan, and Bruce Tabb, eds. 2002. *Water, Leisure and Culture. European Historical Perspectives*. Oxford and New York: Berg.
Andrews, G., S. Chen, and S. Myers. 2014. The 'Taking Place' of Heal and Well-being: Towards Non-Representational Theory. *Social Science and Medicine* 108: 211–222.
Banks, Emily. 2015. Hampstead Heath Ponds Project Is a Dam "Fiasco". *Ham and High*, August.
Beck, Ulrich. 1992. *Risk Society*. London: Sage.
Bennett, Jane. 2010. *Vibrant Matter: A Political Ecology of Things*. Durham, NC: Duke University Press.
Booth, Douglas. 2001. *Australian Beach Cultures: The History of Sun, Sand, and Surf*. London: F. Cass.
Campbell, Agnes. 1918. *Report on Public Baths and Wash-Houses in the United Kingdom*. Edinburgh: Carnegie United Kingdom Trust.
Carter, Simon, Francis Dodsworth, Evelyn Ruppert, and Sophie Watson. 2011. *Thinking Cities through Objects*. CRESC working paper 96. Manchester: CRESC.
Coleman, Tara, and R. Kearns. 2015. The Role of Blue Spaces in Experiencing Place, Aging and Wellbeing: Insights from Waiheke Island, New Zealand. *Health and Place* 35: 206–217.
Corporation of London (COL). 2013. *Hampstead Heath Ponds Project Assessment of Design Flood Summary*, March 2013. Corporation of London.
Davies, Caitlin. 2012. *Downstream: A History and Celebration of Swimming the River Thames*. London: Aurum Press Ltd.
Deakin, Roger. 2017. *Swimming*. London: Vintage Minis.
Digby, Everard. 1587. *De Arte Natandi (The Art of Swimming)*. London.
Foley, Ronan, and T. Kistemann. 2015. Blue Space Geographies: Enabling Health in Place. *Health and Place* 35: 157–165.

Freud, Esther. 2013. Hampstead Women's Pond. *Financial Times*, July 12.
Game, Ann. 1990. Nation and Identity: Bondi. *New Formations* 11: 105–120.
Game, Anne, and Andrew Metcalfe. 2011. My Corner of the World: Bachelard and Bondi Beach. *Emotion, Space and Society* 4: 42–50.
Griswold, A. 1998. *Kenwood Ladies Bathing Pond*. Hampstead: KLBA.
Iveson, Kurt. 2003. Justifying Exclusion: The Politics of Public Space and the Dispute Over Access to McIvers Ladies' Baths, Sydney. *Gender, Place and Culture* 10 (3): 215–228.
Landreth, Jenny. 2017. *Swell: A Waterbiography*. London: Bloomsbury.
Lee, Jessica. 2017. *Turning a Swimming Memoir*. London: Little, Brown.
Lobo, Michele. 2014. Affective Energies: Sensory Bodies on the Beach in Darwin, Australia. *Emotion, Space and Society* 12: 101–109.
Marcus, Helen. 2015. Big Business Will Be Flooded with Cash While the Heath is Mutilated. *Camden New Journal*, February 5.
Marshall, T. 2014. Warning Shot Sent to City of London over Hampstead Heath Ponds Project. *Camden New Journal*, June.
Molotch, Harvey. 2010. Objects in the City. In *The New Blackwell Companion to the City*, ed. G. Bridge and S. Watson. Oxford: Blackwell.
Mukerji, Chandra. 1997. *Territorial Ambitions and the Gardens of Versailles*. Cambridge: Cambridge University Press.
Murphy, C., P. Gardoni, and C. Harris. 2011. Classification and Moral Evaluation of Uncertainties in Engineering Modeling. *Science and Engineering Ethics* 17 (3): 533–570.
Obrador-Pons, Pau. 2007. A Haptic Geography of the Beach: Naked Bodies, Vision and Touch. *Social and Cultural Geography* 8 (1): 123–141.
Orme, Nicholas. 1983. *Early British Swimming, 55 BC-AD 1719: With the First Swimming Treatise in English, 1595*. Exeter: University of Exeter Press.
Osborne, Bruce. 2005. *History of the British Spas Federation*. Spas Federation.
Poynting, Scott. 2006. What Caused the Cronulla Riot? *Race and Class* 48 (1): 185–192.
Rhys-Taylor, Alex. 2013. The Essences of Multiculture: A Sensory Exploration of an Inner-city Street Market. *Identities* 20 (4): 393–406.
Ritvo, Harriet. 2009. *The Dawn of Green: Manchester's Thirlmere and Modern Environmentalism*. Chicago: University of Chicago Press.
Sennett, Richard. 2010. The Public Realm. In *The Blackwell City Reader*, ed. G. Bridge and S. Watson, 2nd ed. Oxford: Blackwell.
Shifrin, Malcolm. 2016. *Victorian Turkish Baths*. Chicago: University of Chicago Press.
Strang, Veronica. 2004. *The Meaning of Water*. Oxford and New York: Berg.

———. 2005a. Taking the Waters: Cosmology and Material Culture in the Appropriation of Water Resources. In *Gender and Development*, ed. A. Coles and T. Wallace, 21–38. Oxford and New York: Berg.

———. 2005b. Common Senses: Water, Sensory Experience and the Generation of Meaning. *Journal of Material Culture* 10 (1): 93–121.

———. 2006. Substantial Connections: Water and Identity in an English Cultural Landscape Worldviews: Global Religions. *Culture, and Ecology* 10 (2): 155–177.

———. 2009. *Gardening the World: Agency, Identity, and the Ownership of Water*. New York and Oxford: Berghahn Publishers.

Switzer, Teresa. 2002. Hungarian Spas. In *Water, Leisure and Culture European Historical Perspectives*, ed. Bruce Tabb and Anderson Susan. Oxford: Berg Publishers.

Thrift, Nigel. 2007. *Non-representational Theory: Space, Politics, Affect*. London: Routledge.

Vogel, David. 2012. *The Politics of Precaution*. Princeton, NJ: Princeton University Press.

Volker, S., and T. Kistemann. 2011. The Impact of Blue Space on Human Health and Well-being—Salutogenetic Health Effects of Inland Surface Waters: A Review. *International Journal of Hygiene and Environmental Health* 214: 449–460.

———. 2015. Developing the Urban Blue: Comparative Health Responses to Blue and Green Urban Open Spaces in Germany. *Health and Place* 35: 196–205.

Watson, Sophie, ed. 2006. *City Publics: The (dis) Enchantments of Urban Encounters*. London: Routledge and Kegan Paul.

———. 2015. Mundane Objects in the City: Laundry Practices and the Making Remaking of Public/Private Sociality and Space in London and New York. *Urban Studies* 52: 876–890.

Weaver, Matthew. 2018. Transgender and the Ponds Study. *Guardian*, June.

Webber, Richard, and Roger Burrows. 2015. Life in an Alpha Territory: Discontinuity and Conflict in an Elite London 'Village'. *Urban Studies* 53 (15): 1–16.

Whatmore, Sarah. 2013. Earthly Powers and Affective Environments: An Ontological Politics of Flood Risk Theory. *Culture and Society* 30 (7/8): 33–50.

Worpole, Kenneth. 2000. *Here Comes the Sun: Architecture and Public Space in Twentieth-Century*. London: Reaktion.

7

Differentiating Water: Cultural Practices and Contestations

Across the world in rural and urban areas, there is a myriad of ways in which people engage with and use water for different cultural practices and different needs, and according to different sets of beliefs, understandings and experiences. Imagine a woman walking across the dry pastures of Senegal in search of a well to fill her bucket, or a boy playing pooh sticks in the muddy canal running through his local neighbourhood in Liverpool, or an old man washing his hair in the Ganges at Varanasi, or a vicar baptizing a baby in an Evangelical church in Melbourne, or a woman contemplating her flooded sitting room in her house in New Orleans after the levees broke in the floods of 2005. Every culture and place has a different relation to this valuable and life-giving resource, mediated by specific histories, engagements, memories and topographies. As hydrologist Richard Meganck puts it (Johnston 2012: vi) 'a river reflects the life and memory of any country and region. Water is mystical, religious, powerful, revered and feared'. Water plays a role in creating and sustaining the diverse socio-cultural relations that characterize everyday life.

Difference constitutes who we are. Where we live, at what time in history, and in what socio-political context has its effects. So too does our gender, age, ethnicity, race, class, sexual orientation, religious affiliation, physical ability and so on. Differences are constituted in relations of power and constantly shifting. The specificity of our entanglements with the material world brings differences into being, which may settle for a while, only to be disrupted later by the rearrangements of the relations and powers from which they emerge. In this complex interplay of differences, water plays a part. Different bodies, subjectivities and groups of people interact with water in different ways. So too the meanings of water, and the rituals and practices associated with it, vary across cultures, people and places. Though difference is an indisputable framework for everyday life, governments, water authorities and managers, typically imagine one sort of user or household and are blind to the needs, desires and practices of the vast diversity of the population they serve. White male subjects, or traditional nuclear family households—at least in the Global North, dominate the agenda—either as providers or as consumers.

Here I aim to depart from this terrain to reflect on water through the lens of two modes of difference—that of religion and, more briefly, gender (in Chap. 3 we saw how the everyday practices of water consumption in the UK and Australia varied across ethnicity and gender). My interest is in the everyday cultural practices associated with water that are played out in city spaces with particular socio-cultural and political effects. Here I want to engage with the different religious uses and practices connected to everyday uses of water that typically go unnoticed, but which are matters of concern for minority populations in cities, particularly Jewish and Muslim communities. Debates around questions of distinct urban cultural practices associated with minority cultures, particularly religious cultures have been played out across the world over at least several decades (arguably since cities began). Particularly contentious has been the wearing of veils and the burka or niqab with different governments adopting different stands, from no restrictions in the public sphere, such as in the UK, to a ban on veils that cover the face in public places in France. One strand of thought advocates the toleration by majority (powerful) cultures of differences enacted in public, while another strand advocates that customs and rituals which differ from the dominant and accepted norms should be consigned to the private sphere. This chapter considers uses and practices of water exercised by minority groups, that are often invisible, but which nevertheless enter the public sphere. Interestingly a search

7 Differentiating Water: Cultural Practices and Contestations

of the major texts on public religion does not refer to them. These are the Muslim practice of Wudu and the Jewish practice of Mikvah in Western cities. Christian baptism is a religious practice associated with water, but one which has its place in churches and, not surprisingly as the dominant religion (at least historically) is not contested or disruptive of cultural norms.

Purification Rituals

> The spirit or breath of God moves on the face of the waters (Genesis 1:2) and sets into motion the divine creation of life over the next 6 days (religion, water in).

In many religions water is associated with purification rituals, seen to wash away sins and impurities (Strang 2001). Water is sublime, a life force, also symbolizing death (Barber 2003: 12), and seen as part of birth and rebirth. Water is central to many creation beliefs, as underground, surrounding the earth or a mysterious abyss from which life emerges. It can signify the Divine itself or nothingness or is seen to have a life-giving power of its own (http://www.waterencyclopedia.com/Po-Re/Religions-Water-in.html.). Water figures in many creation narratives, and even sometimes signifies the divine being, or as life-giving. The idea of a firmament separating upper (heaven) from lower (Earth) waters appears in many religions, where the sky may be a solid dome over the Earth dividing the primeval waters. Those above fall to Earth as rain 'through the windows of Heaven', while the Earth rests on the waters below, from where it originally emerged. Rivers and lakes may separate the world of the living from the dead. Water, flowing downward, symbolizes the transmission of wisdom from on high (ibid.).

Almost all Christian churches or sects have initiation rituals which involve water. Baptism has been a key practice in the Christian church, in its early form as total immersion—a practice still current with the Baptists and Catholics. For example, at the Guardian of the Angels church in the East End of London there is a dramatic looking walk-in marble octagonal font in the middle of the church. More typically, baptism is performed by pouring water over the head of the applicant, often a baby, in a font. The word 'baptism' comes from the Greek word meaning to plunge or to wash and is believed to be the public declaration of belief in Christ and a welcome into the Church, rather than a washing away of sins. Catholics in contrast believe that the stain of original sin is removed. Baptism goes back to the fourth century in the Far East and the fifth century in the West. All churches have a font, and thus a need for a water source. In her article, Oestigaard (2010) focuses on the topography of holy water in England after the Reformation, when the Church set about contesting the belief in the magical powers of water, and the traditional water cult that was seen as a testimony to the power of Satan. Today, more than half of all Christians baptize infants while many others contend that only a believer's baptism and thus adult, is true baptism. Some people insist on submersion or at least partial immersion of the person who is baptized, while others consider that any form of washing by water, as long as the water flows on the head, is adequate. Baptism, unlike Mikvah, is only performed once. Water is also important according to some Christian beliefs for its healing properties, and like other religions is associated with cleansing and purifying the body and spirit, which signifies spiritual rebirth. Lourdes in France is famously a site where the spring water is imagined to heal the sick, which dates back to the mid-1850s when a 14-year-old girl claims to have seen an apparition of the Virgin Mary 18 times. Eventually a church was constructed at the site. Since then, people have claimed to be miraculously healed by the spring waters there. Nevertheless, as already suggested, Christian baptism practices and sites are typically not contested in cities. Rather, some cities where religious water matters owe a lot to the urban and economic energies created by those seeking baptism and cure.

Muslim Practices

> Have not those who disbelieve known that the heavens and the earth were joined together as one united piece, then We parted them? And We have made from water every living thing. Will they not then believe? (Qur'an 21:30)

7 Differentiating Water: Cultural Practices and Contestations

And have you seen the water that you drink? Is it you who brought it down from the clouds, or is it We who bring it down? If We willed, We could make it better, so why are you not grateful? (Qur'an 56:58)

The issue of purification is core to Islamic beliefs, where the Prophet Muhammad emphasizes the importance of cleanliness to Muslims—where 'Cleanliness is half of faith'. In the practices of worship, purity is seen as a pre-requisite to prayer and lack of cleanliness is thought to invalidate prayer. There are two forms of ablution—minor ablution—Wudu, and major ablution—Gushi. The practice of Wudu (Wudhu) entails highly complex instructions as to how it should be enacted, involving different parts of the body, where practices must be performed in an exact order while making the correct supplications. For example, before washing the face it is recommended that believers wash their hands twice, gargle three times and rinse the nose three times. The face must be washed from the upper part of the forehead to the furthest end of the chin and across the breadth of the face to the maximum distance of the tips of the thumb and middle finger. Even if a very small area is left out of the washing, the ablutions will be void. After the face, the hands are washed from the elbows to the tip of the forefingers, followed by the washing of the feet, again in a highly prescribed manner and in the right order. All of these actions must take place before prayer and before touching the Qur'an. The water must be un-mixed and pure although if there is no clean water available, Wudu can be performed with sand or simply by touching the ground and washing with hands (tayammum). There is a fragility to the practice in that it can be broken by going to the toilet, passing wind, touching the opposite sex or a dog, sexual contact. The place where it is performed must also be lawful. A fuller immersion is Gushi which involves washing the whole body as means of purifying oneself from various practices including sexual intercourse, menstruation, post-partum bleeding and 'najasat' caused by touching semen, blood or the dead body (Photos 7.1 and 7.2).

The performance of Wudu in modern city life can be challenging for Muslims, particularly in work places where there is no easy access to water, or where for women, there is a necessity to share intimate spaces

Photo 7.1 Mosque washroom. Sophie Watson

and facilities. Arguably, Wudu reconfigures private/public boundaries and challenges normative notions as to appropriate conduct in public, since in most Western cultures intimate embodied practices are confined to private space, while Wudu brings parts of the body into public places. There are some similarities here with the debates around the wearing of the veil in public (Scott 2010) in that those conventions and rituals that are deemed appropriate are disrupted and contravened. But there are differences also. The niqab is contested on the grounds that the face is invisible which is considered necessary for proper communication in jobs where employees have to deal with the public. In everyday encounters in the street non-Muslims also complain about not being able to engage or make eye contact in the ways that are more typical for urban citizens. The Wudu is potentially problematic since it involves private practices of washing taking place in public, and more particularly the washing of parts of the body that are normally not seen—the feet—at least in the workplace (with the exception of sandals in summer).

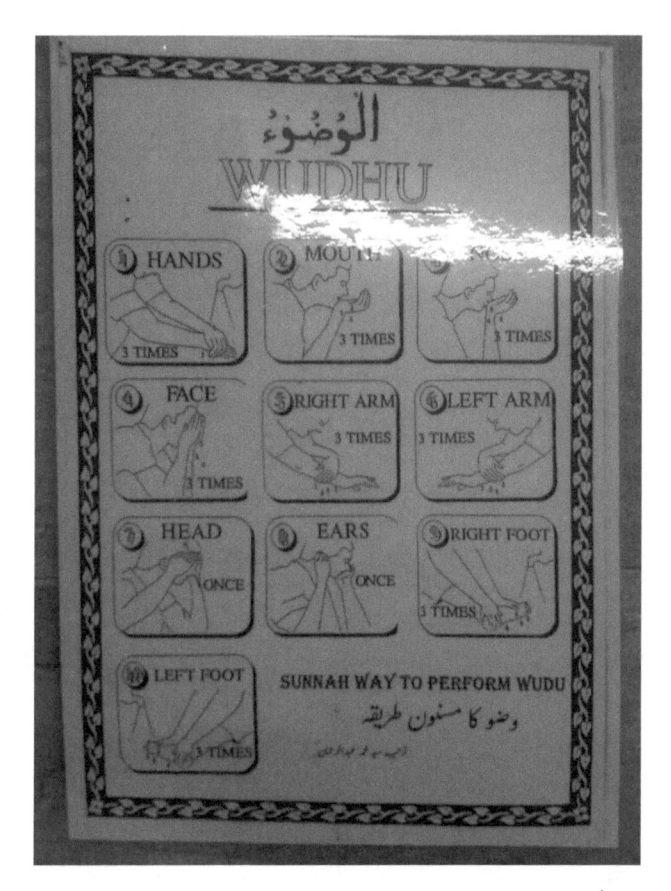

7 Differentiating Water: Cultural Practices and Contestations

Photo 7.2 Wudhu washing instructions. Sophie Watson

We carried out a small study of practising Muslims' engagement with Wudu in Bristol, West England (Watson and Wigley 2018) to consider the implications of the practice in more depth. A number of issues emerged. Cleanliness of the sites to perform Wudu is one concern. Generally, toilet facilities in public places do not have bidets in the UK and urinals are not permitted for use by praying Muslim men as clothes splattered with urine have to be changed before prayer. Some places were described as particularly problematic such as airports and service stations where there is no space for washing. Performing Wudu at work represents a major challenge related to both the facilities and the timing—such as combining Wudu, lunch and prayer in the lunch hour in a location near enough to where people are working. One IT worker described asking to pray in the shower facilities. Strategies to enact Wudu in some form or other include washing feet in the morning and covering them with thick or leather socks which maintains their cleanliness symbolically, or not going to the toilet for the duration of work shifts or travelling. Halal nail polish which is made of water that is permeable is considered possible for the performance of Wudu since unlike other nail polishes on the market, there is no need to remove it before making Wudu (https://www.maya-cosmetics.com).

The chaplain at one university noted that Muslim colleagues were becoming better at articulating their needs to the institution, whereas previously they were 'over-grateful', but others saw special provision as the privileging of one group over another. Many of the problems in non-Islamic countries arise from a lack of understanding of the practice since people in Islamic countries are more used to seeing water spilled on the floor. Co-presence can thus be problematic. Feedback from the wider community included issues associated with practising Wudu in restaurants or hotels, where they encountered non-Muslims who wondered what was taking place. Practitioners described the hazards of spilling water on the floor and trying to balance whilst cleaning feet in the sink. 'We don't particularly want to cause a hazard by splashing water on the floor but it would be nice for that to be incorporated into planning. I think [Muslim] people feel awkward themselves; they don't want to cause a hazard. They just want facilities where they can pray.'

A non-Muslim user of a toilet at the university described seeing people washing their face in the sink, gargling and lots 'of nasal hocking up of… greenies noises which are noticeable when washing my hands'. He was aware that in British culture this is seen as aggressive, rude and intimidating by many and acknowledged that in the context of wider Islamophobia in UK, the provision of Wudu facilities had made him aware of his social conditioning and privilege. A tendency for better practice in larger companies, and obviously in Muslim countries, was reported. For example, one respondent previously worked at a telecom company in Jordan where a special place was allocated for prayer, but there were no facilities for Wudu practice, but as his wife explained 'at least all the people are doing

7 Differentiating Water: Cultural Practices and Contestations

the same so water on the ground is fine'. In his current work (as team leader in WH Smith at a motorway service station) where there were other Muslim employees, his manager previously worked in Saudi Arabia, and thus had knowledge of Wudu and Islam which made him more understanding of people's needs. Yet he reported that washing feet typically causes a mess and a mop and bucket is not always available and can cause hazards or unpleasantness to others. Cleaners at UWE maintained a mop and bucket at all times. An alternative practice (in moderate Islam) is to put some water on the hand and wipe the top of the shoe or a sock, a practice which is not allowed in Saudi/Wahabi traditions. One interviewee explained that in his former country, Jordan, if there is no clean water available then Wudu can be performed with clean sand. This is a wider social practice in Islamic countries where traditionally sand might be used in the place of washing liquid. In this instance water is imagined and symbolic.

Many Muslims reported feeling conscious of the disapproval of non-Muslim others, particularly when Wudu is practised in the sink, and they see water on the floor or someone cleaning/drying the floor. Sutton (2015) in the Daily Mail Australia reported a row that erupted in a Sydney office when a sign was placed on the men's toilet of a Sydney city office block with a cartoon of a man with one of his feet in a sink surrounded by a circle bisected by a slash, in an attempt to ban Muslims from washing their feet in the sink before their daily prayers. A local business student, who worked 20 hours a week in the convenience store, explained to the reporter that two of the five daily prayers required by his religion fell during his 10 am to 4 pm shift at the shop and as was the custom, he washed his hands, his mouth, his head and his neck, as well as his arms, and then placed his feet one at a time into one of the men's rooms' two sinks. Mr Faisal pointed to the floor, saying 'I have done it when you bring water from the sink down to the ground to wash your feet, but it makes too much of a mess, it is not clean and it makes the floor slippery and dangerous.' Like most practising Muslims he didn't feel right if he missed out on any of his five daily prayers, and yet without ablution he could not present himself before Allah to pray, since it is about standing before Allah in a state of spiritual and physical purity. The strata managers of the building supported his view describing the cartoon as 'racist, defamatory and a breach of owner corporation guidelines'. Non-Muslims may feel alienated by the unfamiliar apparatus in Muslim washrooms or processes of washing which can be misinterpreted. Some respondents who could not perform Wudu at work cleaned themselves before work, so they could hold Wudu, but this could mean not going to the toilet, passing wind or touching a woman's hand during the work shift (up to seven hours) in order to maintain Wudu.

Some workplaces are more sensitive to Muslim practices than others. For example, Ian Yemm, the coordinating chaplain at University of Western England in Bristol, told us that washing facilities had been installed in 2010, although initially for men only, but now there are gender separated prayer rooms and washing facilities on all the UWE campuses. These efforts initially proved to be a challenge due to the lack of a market provider of the appropriate apparatus, but more recently, this has been solved by WuduMate which provides footbaths to commercial buildings. Yemm believes that the Estates office are now well versed in Wudu, and that students have shifted from being grateful to being increasingly articulate in communicating their needs as provision has become normalized and expectations are higher. Nevertheless, at one site, Yemm witnessed non-Muslim students laughing at the Wudu sign that looks 'fairly comical' on mixed gender accessible toilets. He expressed concern that this facility should have been used as an opportunity to improve 'religious literacy', although he admitted that this may have been unhelpful or alienating for Muslim users.

Yet Wudu facilities also provide the opportunity for encounters of difference through material practices. Negotiating difference in these sites raises the question of how to balance different needs in the context of a broader sense of lack of resources available for community services. Thus the student union UWE made a proposal for facilities dedicated to Christian use, even though those in existence are underutilized, which the Chaplain explained: 'you didn't have to read between the lines too far to recognise that the reason for the request being made was the idea that Muslims have their own facilities so why can't we have ours?'. Likewise, an Algerian mechanic interviewed outside a mosque following Friday prayers recounted a story from his work where he was raising concerns about health and safety (smoking/loud music) and another colleague said

7 Differentiating Water: Cultural Practices and Contestations

'yeah, there's a lot of things you do that we don't like', in this case meaning prayer practices.

The Prophet and Islam encourage the frugal use of water; it is considered un-Islamic to waste it, but while 1400 years ago Muslims might use a glassful each day, living in the West where water is plentiful leads to waste. Ideally, you should use 500 millilitres to 1 litre of water during Wudu. According to Rizwan Ahmed from the Bristol Muslim Cultural Society, living in western society leads to the challenge of teaching people not to waste water, since in his view none of God's given resources should be treated carelessly. In the local hospice, he has advised providing washing facilities close by so parents are not too far from their children for Wudu particularly during a stressful time. During his father's death, he had to keep Wudu for long times to avoid leaving his father alone too long. Ahmed described an encounter with a local Housing Association who reported to him that one of their families was damaging the bathroom, getting water everywhere and wrecking the tiles. His response was to advise the use of shower curtains and advise the tenants to mop up after themselves.

Respondents described different strategies that they deployed to perform the ritual, for example, we were told that some taxi drivers carry a small canister of water in the boot of the car, so they can perform Wudu sitting on the bumper, and try and find the nearest mosque when needing to pray—the only obligatory prayer is the one performed at sunset. In Muslim countries, most mosques are left open 24/7, while in the west, mosques close 30 minutes after the congregation period to avoid vandalism, with the result that taxi drivers have to go home. Ishan described the challenge of working in Tesco where he used paper towels to dry the facilities raising the curiosity of other colleagues. Unlike in mosques where there is a tap, jug and often a drain on the side of the toilet, no such facilities are provided in non-Muslim spaces. For those who are not flexible, using the sink can be a real challenge. Some men explained how the special leather socks (usually available in Islamic book shops) which allow the feet to move freely so the toes can bend, mitigate the need to wash the feet, since these represent a symbolic cleanse, but summer heat made this practice unpleasant. Interviews with the wider community primarily revealed mystified responses as to what was being enacted in these spaces, but also some understanding of the difficulties Muslim practitioners face: 'I think [Muslim] people feel awkward themselves; they don't want to cause a hazard. They just want facilities where they can pray.'

A minority of non-Muslim urban citizens are aware of what the enactment of purification rituals entails, with the result that Muslim people can feel defensive or embarrassed about performing the practice in public space. A question and answer web conversation recorded in *The Revival*—an online site—reveals interesting further insights from Muslim people. One man raised the question thus:

> Q. I have a room to pray at work but no particular Wudhu (ablution) facilities. I don't feel comfortable washing my feet in the sink or in the toilets at work because I look silly and plus my non-Muslim colleagues look at me funny and say it's unhygienic, so what should I do?

His religious adviser (it is not clear if this is an Imam) replied:

> A. Wudhu is an act of worship and whilst it is obligatory it also carries great reward. Therefore one should not feel silly in taking part in an act of worship. However, there are many organisations who do not allow the washing of the feet in normal sinks due to various reasons. There are also many people who leave a lot of spilt water after they have finished, which gives the practice a bad name. If it is difficult for a person to wash their feet in public places then the following should be done:
>
> As washing the feet is an obligation, the way around this issue is to wear leather socks known as Khuffain. ….If the person takes off the socks once full Wudhu has been done and broken, then they will need to do full Wudhu before they can put the socks back on. Once the socks have been put on they are allowed up to 24 hours (from when the Wudhu first broke) if they are a non traveler and up to 72 hours (from when the Wudhu first broke) if they are a traveler, before they need to take the socks off and wash their feet again.
>
> They are allowed to wear normal socks on top of these special socks but need to wipe on these special socks when performing Wudhu (i.e. take off their normal socks but leave the leather socks on).
>
> Of course Allah knows best.

These comments demonstrate a number of points. First, that this is a ritual that cannot be disobeyed. Second, that there are practical solutions to the problem in introducing new artefacts—the leather sock. And thirdly, there is a recognition that some practitioners need to be more mindful of others in their practice. This latter point is accentuated by others in the same conversation, given a particular emphasis by the only woman contributing to the debate:

Yildiz Yilmaz (23 October 2011).
I'm sorry but I totally disagree. I'm a Muslim woman even I think that looks nasty. I don't want to wash my hands or face after anyone washing their feet in the sink. Let's change how we are as Muslims what do you think? As of today we are dirty, nasty and not hygiene community at all. So just wash your feet at your house in your own SINK!!!

And another participant in the conversation elaborating the notion of rights and the mutuality of respect replied:

We as Muslims must show respect to others and go out of our way to show that we are a civilized people. The condition of bathrooms after some of my Brothers and Sisters have made Wudu is disgusting; water all over the floor, paper towels left on floor, sink counters covered in pools of water. They make no attempt to clean up after themselves. It really speaks volumes to non-Muslims about our character. It is not our right to destroy public restrooms with no regard for others. I have had my dress shirts soiled by getting too close to the aftermath of someone else's "foot bath." If we want respect we must show respect and earn respect. Not act like slobs and call it our religious right.

However, a more common response in the discussion was that people should stop worrying what others think of them, since it is only Allah who ultimately should be the concern:

You need to stop worrying about feeling uncomfortable and feeling silly. Forgot what your non-Muslim colleagues think of you, you are worshipping Allah. So put your feet in the sink and wash it and if anyone comes in and get scared this is an excellent opportunity to do dawah by explaining why you have your foot in the sink (expressing the optimistic belief that) non-Muslims would understand I have been there myself…We can't make our own rules—that wouldn't be Islam.

This adherence to these fixed rules is central to Islam.
And:

> I totally disagree with you, what type of Muslim are you?
> You fear non-Muslims however you don't fear our creator?
> We should be proud to be Muslim and perform wudhu in a public place without fear.
> Don't forget we will not be accompanied by anyone in the grave.
> Let people think whatever but your imam should not be affected.why change for

and: the benefit of others and a detriment to our rules of Islam.

> Oh come on. We Muslims wash our feet round about 5 times a day.

> Our feet can't be as dirty as people's hands which also get washed in the sink.

There are many similar such interventions on the internet, including practical advice and solutions, for example, like the advice on a document entitled 'covertly doing Wudu in a public restroom' which suggests that practitioners should avail themselves of unisex toilets, or bathrooms that aren't commonly used, or bring a friend as support, or take paper towels to clean up and so on. Women's voices are rare however, with the exception of an extended discussion of halal nail polish (Rahman 2017) which is manufactured to enable women to practice Wudu since it lets in air and moisture, which regular nail varnish does not. For Rahman this invention is 'a modern pseudo-science, being used to prop up an idea from pre-history, an idea on its last legs in other cultures, but alive and well in Islam'.

Despite women's relative silence on Wudu, women play an active role in broader water issues. Husna Ahmed, a British born Bangladeshi woman, who is Secretary General at World Muslim Leadership Forum and on the advisory board of East London Mosque, has written a booklet on Islam and Water. In this, she highlights the integral role of water in Islam and its crucial significance for existence and urges fellow Muslims to take water conservation seriously in the context of climate change and new extremes of weather across the world. Part of her strategy is to emphasize the important role of Muslim women through the story of Hajar, the wife of Prophet Ibrahim, who was stranded with her son in the desert with no food and water. In her desperate search, as legend has it, she came upon an angel at the place of Zam Zam who was digging the earth with his wing until water flowed so she could fill her water skin. Until this day, Zam Zam water has remained revered for its heavenly origins and is seen as possessing various health benefits, while Hajar's devotion to her son is upheld as the prime example of exceptional motherhood.

Jewish Practices

Muslims are not alone in seeking some form of purification in relation to prayer. Ritual purification rituals are also important in the Orthodox Jewish religion and are described in the Talmud. Full immersion—tevilah takes place in a Mikvah which is a special bath used for the purpose. Netilat yadayim is the washing of hands. Some ritual washing requires a special body of water and some can be done with tap water. Traditionally the water was required to be from a natural resource, although this rule has been relaxed. Ritual washing is mandatory prior to eating bread with a meal, or after a meal, when getting up in the morning after a full night's sleep, or even after a long nap, or after certain parts of the body—for example, the inside of the nose or ear, or the private parts or arm pits, after cutting one's hair or nails, after sexual intercourse or using the bathroom, after visiting a cemetery or before prayer. Full immersion is essential in a Mikvah following menstruation, the day before important festivals and by some Orthodox Jews, in preparation for Sabbath, or before conversion to Judaism.

Unlike Wudu the requirement to immerse oneself in the Mikvah is not a frequent daily event and thus does not need to be performed in the public spaces of work or college or while travelling. Instead, purification takes place in specially designated spaces—the ritual bath of the Mikvah. In London for example, the Mikvah directory lists Mikvahs in Boreham Wood, Maida Vale, Edgware, Golders Green, Hendon, Ilford, Kingsbury, Stamford Hill and Wimbledon. Some of these are by appointment or open for special days. Mikvahs are of particular significance to women since the ritual washing is connected to women's bodies in particular, due to menstruation.

Feminist scholars have interrogated the ritual of Mikvah arguing that it positions women as a source of moral danger and pollution to men (Hartman and Marmon 2004; Raucher 2017). Mikvah is connected with niddah which defines a woman as impure during her menstruation and for the following seven days. During this time, physical contact with her husband is not permitted. Once her menstruation is over, she is required to go to the Mikvah where she immerses herself for total cleansing—all jewellery is taken off so water can enter every crevice of the body. The action is performed under the supervision of the 'balanit' who certifies that the practice is complete. In her work, Douglas (1970) explored the connections between dirt and cultural impurity and the distinctions made between the impure and the holy, which has some purchase here. Such is the power of the rabbinate that the rabbi is the ultimate judge of the woman's receptivity and if she suspects a suspicious stain, she must show it to the rabbi who will inform her if it is menstrual blood.

In her research Cicurel (2000: 169) asked a rabbi why men should rule on these intimate matters and she was told that 'only men can judge such a delicate matter', which is an astonishing illustration of patriarchal power. Feminist researchers have explored contested meanings and practices around this space of male control. Hartman and Marmon (2004: 396) reported that the Orthodox women they interviewed in Jerusalem felt suffocated by, and were critical of, the requirements of the entire system of niddah and Mikvah. In her research amongst women in Beer Sheeva, Cicurel (2000: 179) found that women contested the practice by simply not attending or reframed and reinterpreted the Mikvah ignoring the label of impurity or turning the discussion of impurity into one of purity. Thus, instead of accepting the notion of punishment delivered to those who refused to carry out Mikvah, the women referred to the associated rewards and benefits. In this way, they took 'religious terms that define them as inferior and reconstruct them in ways that exhibit them as equal if not superior to men'.

Such feminist thought has found its way into material practices and innovations elsewhere. In New York there has been an initiative to reconfigure the space of the Mikvah as a site which embraces a more pluralistic and feminist ethic. *Immersenyc* defines Mikvah immersion as an 'ancient practice that honours the sacred practice of our experiences and our bodies that can mark life transitions such as weddings births. Healing from illness or trauma, significant birthdays, divorce, pregnancy loss, fertility struggles and mourning' What they aim to do is provide a supportive experience where guides are on site to assist people in their experience. These are people from a range of backgrounds and experiences who are open to sharing the experience if so desired. People can attend with friends, or family or alone. According to Rabbi Sara Luria (Lieber 2015) the aim is to imbue the ritual with special meaning and to remove the barriers typically associated with it which act as a deterrent. According to Luria women do not often talk about this highly privatised ritual, so the intention is to enable women to share and speak about their experiences. Through these practices the organisation hopes to remove the negative stigma associated with washing practices. As one of the volunteers put it; My hope is that people will re-imagine the mikvah, see it as a transformative experience, for some people you need something physical to feel spiritual and if the body helps them make sense of the world, mikvah can be used as a tool to do it' (Lieber 2015).

The Mikvah bath has not only been contested by women. Where it has been enacted in a more public space rather than in the more typical effectively private space of the synagogue or other Jewish communal spaces, the practice of Mikvah has mobilized different concerns. In east Jerusalem in the neighbourhood of Silwan, an area seen as critically important for both Arabs and Jews (Medzini 2009), the Pool of Siloam which is a freshwater reservoir, became a source of considerable conflict between Jews and Palestinians in 2009. Here Haredi Jews transmogrified the pool into a Mikvah for symbolic purposes. According to the reporter, on Friday afternoons, dozens of religious and Haredi Jews bearing soap and towels arrived at the pool turning the site into a purification Mikvah. The effect was to exclude entry to women, local Arab residents and tourists. This led to ferocious clashes between Arab youths and the Jews using the pool. From the local Jewish perspective as one young boy said, 'the site belongs to the people of Israel, the people of the Torah and the Bible'. While from the Palestinian viewpoint this practice is considered harassment, which shifts public/private boundaries in a way they contest. As a local resident

Ibrahim put it: 'this is a public place, not a home. Religious people come here, take off their clothes and get in the water. It is a disturbance'. Arguments and outbreaks of violence were common. Mizrahi, cited in the article makes a salient point: 'this is also cited as the original reason that people came to live in the City of David—control over the source of water, which, ultimately is the source of power over the entire region' (Medzini 2009). Arguably then, the pool was being claimed for Mikvah in a move to establish who were the 'masters' of this space.

A Question of Politics

What then does the enactment of purification rituals in urban environments imply for spatial justice? Philosophically Philippopoulos-Mihalopoulos' (2015) argument that spatial justice is the conflict between bodies that are moved by a desire to occupy the same space at the same time frames this dilemma in entirely spatial terms. Such a desire, in Philippopoulos-Mihalopoulos (2015) terms is enacted in the lawscape which is a space of visibilization and invisibilization. In the washroom or public toilet, it is more often the latter, no notices are usually in place, though the Sydney cartoon is an exception, rather, it is a space of unconstrained movement free from legal presence where unwritten codes of behaviour underwrite what takes place. The Mikvah typically is visibly regulated, not so much by law, but by requirements of entry into what is effectively a private (and invisible to the outside) space. Thus, spatial justice is the 'emergence of a negotiation amongst bodies' but this 'is not about equally strong bodies on a flat ontology but about unequal bodies on a tilted surface that by dint of the assemblage of which they are part, they might have an ability to pose the question of emplacement in different parameters than the ones that have thus far determined the power imbalances' (ibid.). In the Mikvah, the inequality of bodies is gendered, though the space is private. But typically, for Muslims practising Wudu in public, there is the necessity to assert their practice in the face of rules and regulations and accepted modes of conduct that have been set by others—implicitly or explicitly. As such, it is an articulation in the context of unequal power relations.

What then, are the political or policy responses entailed in the practice of Wudu in public space? Interestingly in my researches, I have found no mention of Wudu specifically, but it is clearly one aspect of wider questions concerning the governance of religious diversity (Bramadat and Koenig 2009), the separation of religion and the state, notions of tolerance and the place of religion in modern liberal democracies. As Bramadat points out, given the rise of secularism in some countries and the idea that the place of Christianity in society has been settled decades ago, for many people the debate over the accommodation of religions brought by new arrivals in established cultural structures seems regressive, especially when this means having to accept that the surfaces of shared washing spaces may be wet. The contentious issue regarding the public expression of religion in recent years has been around the wearing of the veil, and even more so around the hijab or niqab. But even symbolic expressions of religious practice which are far less obviously intrusive than Wudu practices, such as the siting of the eruv boundary (which redefines public spaces as private for the purposes of the Sabbath) in Orthodox Jewish communities (Watson 2005) or 'unusual cultural practices' in the public celebration of yoga (Bramadat 2018), can come to be fiercely contested. Countries have taken different stances on the separation of church and state, with France notably adopting the principle of *laïcité* which sets up a clear delineation between the two spheres, and which relegates religious practices like the wearing of the veil to the private sphere. That said, this distinction is fuzzy in practice, since old Roman Catholic sites still benefit from state support, and some religious practices like the ringing of bells are condoned, while others are not—such as the call to prayer.

In Canada, the idea of state neutrality, as defined by the Supreme Court, is assured when 'the state neither favours nor hinders any particular religious belief, that is, when it shows respect for all postures towards religion, including that of having no religious belief whatsoever' (Berger and Moon 2016: 4).

Nevertheless, as Moon and Berger suggest, this notion is not without its limits, conundrums and frailties, such as in trying to determine whether the use of religious symbols is an acknowledgement of that country's religious history or whether it amounts to a clear affirmation of the truth of a particular religious system. This will be even more the case where there is a majority religious culture, and particular festivals are chosen for annual holidays—for example, Christmas and Easter. Wearing religious symbols at work represent a terrain closer to the one discussed in this chapter which has a bearing on this discussion of neutrality. On this, Maclure (2016: 21) while acknowledging that the principle of state neutrality is compatible with the banning of religious clothing or displaying religious symbols in public buildings, usefully argues that though citizens seeking public service have the right to be treated in a religiously neutral way by public sector employees, they do not have the right to deny employees signs of their religious affiliation. More generally, she suggests that tolerance involves bearing non-harmful behaviours even if we find them objectionable or irritating.

Thus, the acceptance of others washing in public implies a tolerance of behaviours that are different from the majority population. This is not a question of assimilation into dominant norms, where one group has the power to accept and tolerate, or not, the other (Hage 2012). Rather, since the practice is central in the everyday lives of some Muslims in the city, to refuse its enactment is to refuse the different culture and religion of large numbers of people who live in many Western cities. To regulate or ban the practice to the private sphere is to ban those that practice their faith also, since for many Muslims the requirement to purify before prayer is seen as crucial to their religion, though there is clearly a diversity of perspectives and enabling practices as the research cited here made clear. Given there are few public places which are designated as specifically for the practice of Wudu, everyday practices of washing will continue to be enacted in shared space requiring negotiation and recognition between practitioners and observers in the context of unequal power relations. Such negotiation can take different forms.

The Mikvah raises different spatial questions. In the case of Pool of Siloam in Jerusalem, the occupation of the pool by one group—the Jewish men performing purification rituals in this public space, necessarily turns the space into a space of exclusion—exclusion of Palestinians, women, tourists and others. Unlike the purification of Sikhs in the lake at the Golden Temple, which takes place as strangers walk past and share the space, the Pool of Siloam becomes a space of enclosure with fixed and impermeable boundaries (Sennett 2010) where power is exercised through the enactment of rituals. Contestation is thus inevitable and not clearly open to resolution. The contestation of women to the Mikvah involves a different set of strategies. Here women are resisting the control of men over their bodies and embodied rituals. From the evidence reviewed here, there are two distinct and interconnected strategies. One is to redefine the meanings of the Mikvah, by reversing dominant understandings of the rituals and re-inscribing them with others. The second is the creation of alternative spaces of enactment where new meanings and experiences of the purification rituals are constituted and performed thus freeing the Mikvah from what some women see as an oppressive set of prescriptions.

Votive Objects and Scattering of Ashes

Rivers are not only imbued with notions of purification, they are also significant sites for the deposit of votive objects and for the scattering of ashes. In the Hindu religion, the Ganges is seen as the most sacred of rivers where people bathe to pay homage to their ancestors and their gods, cupping the water in their hands, lifting it and letting it fall back into the river. Many Hindus desire to have their ashes scattered in the Ganges, including Nehru who wrote in this will: 'The Ganga is the river of India, beloved of her people, round which are intertwined her racial memories, her hopes and fears, her songs of triumph, her victories and her defeats. She has been a symbol of India's age-long culture and civilization, ever-changing, ever-flowing, and yet ever the same Ganga.' (Eck 1982: 214–15). The practice of depositing objects dates back to the Bronze and Iron Age objects which Richard Bradley (1990) concluded were part of religious rites carried out in honour of deities inhabiting the waters (Photo 7.3).

The practice appears to have been common across several religions. Gould (2005) reports how archaeologists from the Museum of London have been puzzling over seemingly well-preserved finds such as urns, wall plaques and statuettes of Hindu gods found along the foreshore. According to Hindu priests, these latest artefacts are either ceremonial water carriers used in purification ceremonies or containers for the ashes

Photo 7.3 The river Ganges at Varanasi. Jessie Watson

of dead relatives. Soapstone and metal statuettes of the elephant god Ganesha and the monkey god Hanuman have been washed up from Bankside in the City right down river to the East End. Other objects include ghee lamps used during recent Diwali celebrations and an intricately painted copper Yantra plaque—a talisman to ward off evil spirits. According to Gould, there are different interpretations as to their significance. Faye Simpson, community archaeologist at the Museum of London, believes the findings, were either placed in the Thames in the hope that they would find their way back to the source waters of the Ganges, "or more likely the Thames has become a surrogate for the Ganges and has a religious significance of its own, and part of the spiritual life of Hindu communities", while Ramesh Kallidai, secretary general of the Hindu Forum of Great Britain suggests that once a household

7 Differentiating Water: Cultural Practices and Contestations

deity gets chipped or broken it cannot be used for worship and must be buried, burned or immersed in water (Gould 2005). Mudlarks who retrieve objects from the Thames report finding many such religious artefacts. Included in their finds, as well as Hindu objects, are Pewter pilgrim badges, thirteenth century keepsakes gathered by god-fearing ancestors from religious shrines across the country.

The scattering of ashes is a practice enacted by many religions which in London has been commercialized in the form of a small family business. The various options include firing the ashes from a WW11 plane, or the possibility for the bereaved of taking a boat down the Thames from Lambeth Bridge to scatter the ashes of loved ones in a specially designed biodegradable urn to prevent them being blown away, with flower petals for scattering and dissolvable notelets for promises offered to enhance the experience. So popular is the practice that the Port of London Authority has had to produce special guidelines. As they put it:

> the loss of a loved one brings about a difficult season of life that takes time to adjust and heal from the pain. If you have decided to have the one who has passed on cremated and wish to a scattering ashes ceremony over the tidal Thames in England, there are some things to keep in mind. These include not holding ceremonies on the bridge for fear the ashes will reach passing boats, avoiding windy days, and not leaving objects that are not biodegradable.

They even deploy a language of affect in the material: 'Putting together a simple time of reflection where you and your party board the boat to scatter ashes and say a few words of remembrance will go a long way in the mourning process'.

Deploying Religion for Water Conservation

The centrality of water to many concepts and practices of religion, lends it an intriguing power in persuading people to shift habits and behaviours in relation to the careless use of water, which characterizes consumption in many cities of the world, where water appears to run freely from the tap as a boundless source. Yet as I have argued, even in countries with a relatively abundant water supply, in the context of climate change, water is likely to become an ever-scarcer resource. As we saw in Chap. 3, many water authorities now engage in strategies to lower consumption and change water behaviours. In her booklet on Water and Islam, Husna Ahmed goes to considerable length emphasizing the significance of water for survival and the need to preserve it for the benefit of future generations, taking issue with technocratic solutions for conservation and exhorting 'you and me' to change our ways. She emphasizes the growing concerns about the availability of safe water as the global population increases, emphasizing that more than half of children in the world live in households without access to improved water and sanitation, which puts their survival and development at grave risk (1.5 million children under five die every year because of diarrheal diseases alone). In her booklet, she quotes the Prophet Muhammad: 'Muslim Ummah is like one body. If the eye is in pain then the whole body is in pain and if the head is in pain then the whole body is in pain' and returns to the story of Hajjar to emphasize the special role mothers have to publicize this message. There is an extensive set of recommendations for Muslims in developed and less developed countries. For example, in the former she advises the practice of Wudu in a bucket, the substitution of tap water for bottled water, not leaving the tap running when tooth brushing or washing up, taking shorter showers and installing water saving devices and recycling water—in other words reiterating the messages promoted by water authorities like Thames Water. For those in developing countries she advises Rain water harvesting, storage and water purification, treating water to make it safe to drink, and planting trees. Large religious organizations like the World Council of Churches' "Justice, Peace, Creation" have initiated a campaign which is particularly concerned with ecology and climate change. As people face the challenge of sustaining the world's water today and for the future, religion plays an increasingly recognized ethical and practical role in water conservation.

In their *Love Every Drop* campaign to reduce water consumption Anglian Water worked with Islam Peterborough to take their messages to the Muslim community, drawing on the high value attached to water and its place in maintaining purity and cleanliness in Islamic faith. Imams from Islam Peterborough spoke to over 3000 people at Friday prayers about the importance of keeping water clean and explained how residents could act to reduce contamination in the water in their own communities. At the same time Islam Peterborough's women's groups were encouraged to put waste like cooking fat, baby wipes and other items into the bin rather than down the sink or toilet into the water supply. According to the water authority, this strategy contributed to the 60% reduction in water blockages in just a few weeks. Muslim initiatives to curtail water use are not alone. In July 2017, the Vatican turned off 100 fountains in Rome in response to the prolonged drought that had beset this small city state and the city of Rome through the early summer, following lower than average rainfall. In this, they were following the lead taken by the city to ration water supplies. This was the first time that authorities in the spiritual home of the world's 1.2 billion Catholics could remember being forced to turn off the fountains. The decision was wholeheartedly supported by Pope Francis who sees caring for the planet and its resources as an important issue, reflected in the first papal document related to the environment and ecology.

Gendering of Water

> How do we recognize the centrality of gender as an organizing principle in the ways water is envisioned, used, managed and interacted with in everyday lives, at different locations and in different contexts? (Lahiri-Dutt 2006)

A sex/gender lens throws up rather different questions and issues from those related to religious rituals of purification. There are however similar things at stake: that is, they both contribute non-dominant forms of knowledge, and they both are productive of difference and inequality. As such, I am considering them both here. Sex/gender differences are constituted in relations that are contextual, historically, culturally and temporally specific, not fixed but shifting, and mediated by a complexity of material and socio-cultural factors that cannot easily be read off by any fixed or pre-determined assumptions. Sex/gender differences produce other differences which in turn mould and re-mould the very differences that produce them. In this way of thinking water too is a material substance which mediates and constitutes gender relations in place. Naguib (2009) for example, describes the much anticipated and desired arrival of water in 1985 at Musharafah—a village on the West bank, after years of long journeys to collect water from the spring. Instead of the imagined fair distribution, greater sanitation and easier irrigation, women describe the loss of physical place of social encounters, and conversations. Water, though often seen as formless and fluid, often reflects broader social and cultural values and inequities in access to social, economic and political power, and these are reflected in water use and decision making.

In many instances where there is an inadequate supply of safe water, women bear the brunt of this. The different ways in which water scarcity or water abundance impacts on women is contingent on the wider social, cultural and economic world in which it takes place. We saw in Chap. 3, how dominant definitions of women's beauty and expectations around notions of women's bodies and femininity in the British context produced practices of long showers particularly amongst younger women. In countries where water is scarce, and where women are defined as those primarily responsible for the maintenance of domestic life, then it is women who make the long journeys each day to find enough water to supply their families. It is women who bear the greatest burden of water provision and who are primarily responsible for water sanitation and maintaining the health of their families. Women are deeply embedded in the very stuff of water on an everyday basis. Daily practices associated with water for women living in a poor shanty town on the edge of Rio de Janeiro bear little resemblance to the daily practices of a woman living in Manhattan. They are worlds apart.

Yet, though women in many parts of the world have an intimate knowledge of water, their knowledges remain untapped and decisions as to its provision and use at community, local and national levels are made elsewhere and subject to masculine control. Water management processes are driven by 'the hydraulic mission' (Earle and Bazilli 2013), where technical and engineering solutions dominate resource management, where 'big water'—as Allon and Sofoulis (2006) call it—is the dominant discourse—one which excludes on the ground messy and day to day knowledge and experience. Where water is viewed as an 'economic good', the emphasis shifts to questions of cost and profit, and again denies the place of water in health and sanitation where women have greater interests. What passes as gender neutrality in the technical solutions and provision of water in reality masks gender biases and assumptions in a multitude of ways. Women are underrepresented in decision making bodies.

Fluid Bonds (Lahiri-Dutt 2006) explores how decisions about the use and management of water in Australia can benefit from the inclusion of a gender perspective. At the policy level, it advocates the participation of women in decision making processes where water is involved, drawing attention to the need to investigate who actually benefits from different policies. The report recognizes that the complexity of water-related resource and policy decisions formed in a broader frame of a technically driven engineering system does not fully appreciate and incorporate the different social relationships and roles of women and men, poor and rich communities and minority and majority cultures. Gender issues in water are varied and are necessarily imbued with dominant gender norms and relations, which often go unnoticed, and which ignore differences between cultures, where minority cultural practices have the potential to be subsumed and forgotten. Instead, the report advocates a more nuanced understanding of difference, where gender is recognized but also inflected with diversities of class, age, ethnicity and location. As Lahiri-Dutt argues, just as water is fluid, so too are the social constructs which shape women's lives. What may appear as gender neutral is often intertwined with cultural values.

Feminist scholars have contributed a different way of conceptualizing women and water that shifts the focus from the pragmatic and political terrain. Here contemporary philosophical and psychoanalytic thought proposes new subjectivities that are watery and watered. Neimanis (2013) is one such thinker who proposes the figuration of the body and water, where the post-humanist and material subject is attuned to other watery bodies within global flows of political, social, cultural, economic and colonial power. Pointing out that water gives us life, and that we are all bodies of water, opens up the possibilities for a post-humanist feminism which addresses the need to cultivate more ecologically responsible relations to water (ibid.: 24), where we 'each chart our politics of location in a way that recognizes our diverse aqueous implications and responsibilities', where we are more attuned to more than human bodies.. Other scholars have connected insights from feminism to indigenous people or questions of migration (Perera 2013). Thus, Wong (2013) reflects on 'the Healing Walk' on the tar sands of Alberta, Canada, where First Nations communities from all over the world gather to call for the protection of 'water, air and land for future generations'. Wong here emphasizes the crucial significance of clean water in everyone's lives and the ways in which industrial modernity and the capitalist profit motive have eroded the common inheritance of all humanity, but especially that of the indigenous communities whose land, water and air are being polluted, appropriated and destroyed in order for oil companies to extract petroleum from tar sands.

Conclusion

Differences in the city constitute, and are constituted by, different urban sites and spaces. Differences are produced through social, cultural and material relations, and embedded in them, often in unpredictable and unexpected ways. In post-humanist thinking the material subject adapts to other watery bodies within global flows of political, social, cultural, economic and colonial power. This is one way of thinking about differences. Other frames are more attuned to the structural systems in which differences become manifest. In this chapter, we saw how purification rituals using water make certain kinds of differences, in this case, religious differences, visible in ways that are not typically seen. In other chapters in this book, differences are produced through immersion in water for pleasure, or through strategies of everyday domestic consumption. Though differences are often overlooked in discussions of everyday life, in urban policies, in politics, amongst many other spheres, thinking about water as undifferentiated in its flows through the city is to overlook an important way in which water matters.

References

Allon, Fiona, and Zoe Sofoulis. 2006. Everyday Water: Cultures in Transition. *Australian Geographer* 37 (1): 45–55.

Barber, T. 2003. *Waterfalls and Fountains*. Neptune City, NJ: TFH.
Berger, Benjamin, and Richard Moon, eds. 2016. *Religions and the Exercise of Public Authority*. Oxford and Portland, OR: Hart Publishing.
Bradley, Richard. 1990. *Passage of Arms*. Quoted in http://theses.univ-lyon2.fr/documents/getpart.php?id=lyon2.2009.beck_n&part=159203.
Bramadat, Paul. 2018. *A Bridge Too Far: Understanding a Scandal in the Pacific Northwest*. Working Paper.
Bramadat, Paul, and Matthias Koenig. 2009. *International Migration and the Governance of Religious Diversity*. Montreal and Kingston; London: McGill-Queen's University Press.
Cicurel, Inbal. 2000. The Rabbinate Versus Israeli (Jewish) Women: The Mikvah as a Contested Doman. *Nashim: A Journal of Jewish Women's Studies and Gender Issues* 3: 164–190.
Douglas, Mary. 1970. *Purity and Danger: An Analysis of Concepts of Pollution and Taboo*. London: Routledge and Kegan Paul.
Earle, Anton, and Susan Bazilli. 2013. A Gendered Critique of Transboundary Water Management. *Feminist Review* 103: 99–119.
Eck, Diana. 1982. *Banaras, City of Light*. New York: Columbia University.
Gould, Mark. 2005. Riddle of the Hindu Relics. *The Guardian*, November 1.
Hage, Ghassan. 2012. *White Nation: Fantasies of White Supremacy in a Multicultural Society*. London and Sydney: Taylor Francis.
Hartman, Tova, and Naomi Marmon. 2004. Lived Regulations, Systemic Attributions: Menstrual Separation and Ritual Immersion in the Experience of Orthodox Jewish Women. *Gender and Society* 18 (3): 389–408.
Johnston, Barbara Rose. 2012. *Water, Cultural Diversity, and Global Environmental Change Emerging Trends, Sustainable Futures?* Paris: UNESCO.
Lahiri-Dutt, Kuntala, ed. 2006. *Fluid Bonds: Views on Gender and Water*. Kolkata: Stree.
Lieber, Chavie. 2015. A New Kind of Mikveh in New York City. *Tablet*. Online. https://www.tabletmag.com/scroll/147635/a-new-kind-of-mikveh-in-new-york-city.
Maclure, Jocelyn. 2016. The Meaning and Entailment of the Religious Neutrality of the State: The Case of Public Employees. In *Religion and the Exercise of Public Authority*, ed. D. Moon and B.L. Berger, 11–23. Oxford and Portland, OR: Hart Publishing.
Medzini, Ronen. 2009. Religious Use Contested East J'lem Site as Mikvah. *Jewish World*. https://www.ynetnews.com.
Naguib, Nefissa. 2009. *Women Water and Memory: Recasting Lives in Palestine*. Leiden and Boston: Brill.

Neimanis, Astrida. 2013. Feminist Subjectivity, Watered. *Feminist Review* 103: 23–41.

Oestigaard, Terje. 2010. The Topography of Holy Water in England after the Reformation. In *Perceptions of Water in Britain from Early Modern Times to the Present: An Introduction*, ed. Karen V. Lykke Syse and Terje Oestigaard, 15–35. University of Bergen; Bric Press.

Perera, Suvendrini. 2013. Oceanic Corpo-Geographies, Refugee Bodies and the Making and Unmaking of Waters. *Feminist Review* 103: 58–79.

Philippopoulos-Mihalopoulos, Andreas. 2015. *Spatial Justice: Body, Lawscape, Atmosphere*. London: Routledge.

Rahman, Simi. 2017. https://medium.com/@outspoken/the-problematic-politics-of-halal-nail-polish-or-how-pseudoscience-is-being-placed-in-the-service-509d5a8f1162.

Raucher, Michal. 2017. Immersing in Climate Change. *Journal of Feminist Studies in Religion* 33 (3): 162–167.

Scott, Joan. 2010. *The Politics of the Veil*. Princeton University Press.

Sennett, Richard. 2010. The Open City *Quant*. In *The New Blackwell Companion to the City*, ed. Gary Bridge and Sophie Watson. Oxford: Wiley.

Strang, Veronica. 2001. *Evaluating Water Cultural Beliefs and Values about Water Quality, Use and Conservation*. Ipswich: Water UK Publications.

Sutton, Candace. 2015. It's About Being Respectful to My Religion: Storm of Controversy over Sign Banning Muslim Staff from Washing Their Feet in City Office Public Toilet before Their Daily Prayers. http://www.dailymail.co.uk/news/article-3071556/Muslim-staff-outraged-mystery-office-worker-posts-sign-banning-washing-feet-hand-basin-daily-prayers.html.

Watson, Sophie. 2005. Symbolic Spaces of Difference: Contesting the Eruv in London and Tenafly. *Environment and Planning D: Society and Space* 23 (4): 597–613.

Watson, Sophie, and Edward Wigley. 2018. *Religious Rituals in Public Space*. Unpublished Research Paper, Open University.

Wong, Rita. 2013. Ethical Waters: Reflections on the Healing Walk in the Tar Sands. *Feminist Review* 103: 133–139.

8

Water Traces in Urban Space

Cities are alive with fragments and traces of the past. Every street in the city is a palimpsest of other times, different urban strategies, different needs, representations and intentions. Each material object is embedded in a complex network of the socio-technical relations and politics at the time of its making. The making and remaking of objects in place, their destruction, neglect on the one hand or their valorization as heritage objects or things to be preserved and saved, and sometimes still used, tell us stories about the different periods through which they have travelled, or remained static, become visible or invisible.

Many such material artefacts or traces are barely remarked, part of an everyday life that has long gone or been forgotten. Others are brought into visibility by the interest or concern they excite in planners, historians, conservationists, heritage makers and others who inhabit the city. These are the objects that are saved from neglect, turned into history, standing in for days that are long since passed, speaking for, and representing, those who no longer can speak, and opening up memories and new perspectives. As city dwellers we walk through city streets and spaces often oblivious to the objects around us that have become so familiar as to render them out of sight, unremarkable and part of the urban fabric that we take them for granted as we meander past or rush to the bus, to work, or to meet a friend. And for each and every one of us they have different meanings and connections.

Nostalgia plays its part in this story, where certain material forms represent a way of life and traditions which are now revered and even mourned. Some writers (Edensor 2005) have pointed out the haunting effects of the material traces that remain part of the urban fabric. Struggles over the preservation or destruction of one object over another inevitably are implicated in relations of power involving inclusions and exclusions, where dominant voices define their representation. Objects are active and vibrant players in the city assembling different publics in new connections and alliances. Sometimes these publics gather together unlikely bedfellows connected only by their desire to preserve a site or artefact, or by their opposition to a development, or the destruction of a monument, or by protection of the environment.

Different interests and publics in the city are not stable but are constituted through their acts of representation or action. Walter Benjamin is one such writer who argues that all writing of history represents dominant bourgeois values, suggesting that random objects should be allowed to collide with one another creating new phantasmagorias of city experience for the urban flaneur. For Benjamin (1982) the Paris arcades are the quintessential space of consumption in the late nineteenth and early twentieth century. Deploying heritage as a strategy for urban regeneration has come in for considerable critique for being staged and packaged such that history is only allowed to be experienced in a commodified form, and for telling an incomplete story where only certain practices become reified in monumental geographic sites (Hetherington 2014). And in the complexity of cities and their layered urban textures and fabric, different histories sit side by side opening up the past as a multiplicity of moments and forms.

Water is one key player in this making of city materialities. As a vibrant substance coursing through cities in rivers and canals and man-made waterways, it brings into being a plurality of material forms which speak to the histories of the times of their making. Many such watery objects and artefacts no longer have a purpose and are out of sight, no longer of use in the everyday lives of highly globalized and technological environments that characterize contemporary cities. The socio-technical configurations within which they were first situated have shifted, as rivers become less central to city economies, as sewage is processed in new ways, as socio-cultural practices of consumption change and so on. But anyone who takes a walk through the city mindful of water and the spaces, objects and artefacts that it has made at different moments in histories will see these water traces and fragments. Sometimes they are valorized and remade as heritage sites, sometimes they are reconstituted for different uses, and sometimes they are neglected and forgotten—only visible to the discerning eye. Through the photographs in this chapter, I hope to show how water has made its many marks in the city, often forgotten but with an often-vibrant presence nevertheless.

Horse Troughs

Walking through the city streets of London and Sydney a chance encounter with a horse trough recalls a time when horses were numerous in the city, either ridden alone or pulling carts or drawing a passenger omnibus. Sometimes they also were used for cattle as they were herded to the market for sale. In London the Metropolitan Free fountain drinking association, established in 1859 to provide fresh drinking water, was expanded in 1867—including horse troughs in its name—to include the provision of troughs for animal welfare. By the early twentieth century it was estimated that one trough provided water for 1800 horses in any one day (Photo 8.1).

In Australia one George Bills from Brighton, who emigrated to Sydney in 1885, established a business to manufacture bird cages, door mats and mattresses. George and his wife Annie were keen supporters of animal welfare charities like the Society for the Prevention of Cruelty to Animals, and on his death his Estate set up a Trust to erect drinking troughs in the city, which were to be inscribed with the names of George and Annie—subsequently in true sexist fashion, to be known as Bills Horse Troughs. In 1931, for example, eight troughs were constructed in Victoria at a cost of £13 each. A recent enthusiast, George Gemmill (see billstroughs@hotmail.com), has identified 200 locations in Australia where these troughs can be found. In Sydney, as in London, many of these remain abandoned and empty, serving as a reminder to a different era before the rise of the motorcar.

Photo 8.1 A horse trough in Sydney. Sophie Watson

Water Tap, Africa

The water tap is a ubiquitous material object in countries of the Global South, sometimes fallen into disuse and sometimes an essential part of the contemporary water supply, but never an object revivified as heritage. The water pipe standing alone in the city street or rural village is a reminder sign of the limited connection of private dwellings in the poorer areas to a municipal water supply. In the socio-technical landscape of everyday life, it serves as a reminder of the gulf between the private worlds of ever-flowing clean water, inexpensive to acquire and a taken for granted facet of everyday life at the turn of the tap in the dwellings of the Global North, and the scarcity of water in many parts of the world, where people may walk many miles to serve their daily needs (Photo 8.2).

The water tap is not easily classified. As de Laet and Mol (2000) argue though, it is solid and mechanical rarely can it be clearly said to work or not work. Rather it is a 'device installed by the community, a health building promoter and a nation building apparatus' which may

Photo 8.2 An African water tap

work as a water provider but may not bring health—it may work in the dry season, but not the rainy season, or may work for a while then break down.

A Mail Online article asserted that tap water in every African country is 'unsafe for human consumption', a statement that was quickly dismissed as a gross generalization. Accurate figures are hard to establish and water quality between countries and at any one time vary across regions and time. Wilkinson found that in Ethiopia, out of 832 piped-water samples 80.4% met the World Health Organisation guideline standards for micro-organisms, compared to 44% in Nigeria and 98% in South African municipalities.

Where the water tap runs dry it is a stark indication of the increasing levels of drought faced in large parts of the world.

Thames Steps

The Wapping Old Stairs drop down to the river Thames very close to the Prospect pub (dating from 1520). A witty local erected a noose in honour of Judge Jeffries known as the 'Hanging' Judge Jeffries who was a regular drinker at the Prospect. At one time there were some hundreds of stairs down to the Thames. By 2018 there is no clear estimate as to how many remain in existence, some reports claiming only a dozen or so remain, though this is a considerable underestimate according to other sources. Some of the old stairs are preserved while others have fallen in a state of decay (Photo 8.3).

These steps known as the Watermen's Stairs dropped from street to the littoral water level and were used at least from the fourteenth century as access points to the watermen who took passengers up and down and across the river. Part of the apprenticeship of the watermen (still in existence, see Chap. 4) was to memorize the location of the stairs, many of which were built close to a public house—useful in the event of a drunk falling into the water after a boozy night. Before the construction of the bridges, when only the Old London Bridge was in existence, alleyways

Photo 8.3 The Wapping Old Stairs

leading down to the boats were the only way to cross the river. In the sixteenth century following a risky spectacle involving the Royal Watermen shooting the rapids, Henry VIII decided that the river needed cleaning up, which involved the removal of the wooden structures and water wheels built into the river, and their replacement by a series of permanent walls and stairs down to the water's edge. In 1746 the surveyor and cartographer John Rocque published a map of London that listed over 100 stairs and docks which were in use on the Thames at the time. During the era of steamboats during the mid-nineteenth century the steps were widely used, but by the later part of the century the increased use of the Hackney carriage led to their increasing demise, followed by their complete decline as the river ceased to be central to London's transport network. A text from the mid-1920s—Wonderful London—depicts two sets of the old Watermen's Stairs on the Thames, where under the picture of the Ratcliffe Cross Stairs the caption reads: '[A]n ancient and much used landing place and point of departure of a ferry. There is a tradition that Sir Martin Frobisher took boat here for his ship when starting on his voyage to find the North-West Passage.'

The bombing of the docks in the Second World War, combined with the decline of the central London docks in the 1970s/1980s as container ships which made the delivery of freight up the river redundant, opened up a space for the property development industry to redevelop the old small docks, and pubs, and the steps nearby, into expensive residential and commercial sites. After a couple of decades of neglect once again attention was drawn to the state of the steps in the Port of London Authority (1995) report 'Steps, Stairs and Landing Places on the Tidal Thames' which contained the first comprehensive survey of their condition, status and use. By the turn of the twenty-first century, once again these traces of a vibrant water inflected material past were reinvigorated by local history and urban activist groups. Like other objects depicted here activists argued for the recognition of significant old sites and a re-valuing of the Thames heritage, where the stairs represented an important route into accessing the riverside for local communities. As a result, in 2002, The City of London recommended that river and foreshore access, the stairs and steps on the Thames, should be opened up where there is a 'historical precedent and a practical need'. Eight years later member of the River Thames Society's Central Tideway and Estuary Branch Peter Finch compiled a similar survey to chart the previous 15 years since the PLA report, detailing their current condition and accessibility. Though their presence is largely hidden from visibility, these steps are a reminder of an earlier time when water was central to everyday life in the city.

Rye Cistern

In different cultures at different times rainwater has been collected in various receptacles. A cistern is one of these. In many parts of the world where water was scarce, cisterns were often constructed from waterproof lime plaster and located in the floors of houses or castles. Water was used for washing, irrigation and cooking. In recent years their use has largely been confined to irrigation due to concerns over water purity. In some Southeast Asian countries, where pouring water over the body by a receptacle is the traditional showering practice, houses are constructed with a cistern to retain water for bathing.

In the UK cisterns are no longer used for the provision of public water, but their material traces remain in towns and cities. In Rye in East Sussex, where the water system dates back to the sixteenth century, a brick cistern survives in the corner of St Mary's churchyard as witness to the town's early water system, predating more well-known Victorian water systems by a couple of centuries. Rye, now known largely as a popular tourist destination, has a significant history as a trading port, forming one of the confederations of historic coast towns known as the Cinque Ports. The safe harbour of the town was gradually lost over a century ago lost through long shore drift and storms resulting in a two-mile separation of the town from the coast.

In the sixteenth century Rye's mercantile success brought prosperity to the town which translated into growing civic infrastructure for the increasing population including a number of water cisterns (Photo 8.4). According to early records the pipework received regular attention, with repair contracts issued between the Mayor and his municipal officers with various artisans to maintain, mend and solder the pipes or relay new ones. As

Photo 8.4 A cistern in Rye, Sussex. Sophie Watson

early as 1684, and again in 1701, local taxes were levied for the repairs to be carried out. In 1727, Rye Corporation appointed a committee for water supply, which was estimated at £600. In 1733, the 19-mm bore lead pipes laid originally were removed because they were 'too small to convey the water from the new conduit to the Strand' and replaced by 51-mm bore elm pipes, which enabled water to be carried to the highest point in the town at St Mary's churchyard, where a red brick cistern and tower were built between October 1733 and 1735 with a capacity for 90,920 litres of water at a maximum depth of 2.4 metres. In 1754, it was reported that calves' feet had been found in the reservoir, poisoning the water, leading to a local edict which stated that anyone who deposited 'dirt, dust, soil, trash, nastiness or anything else' in the cistern would be liable to prosecution.

By the late nineteenth century, the cistern had fallen into disuse due to the construction of waterworks in Playden, north of Rye. Like many such artefacts of water infrastructures the cistern has been reconstituted as a heritage site, where it has been designated as a Grade 11 listed building and a Scheduled Ancient Monument. And like many such material traces of different socio-technical networks, the cistern has become the site of contestation. In this case a local resident was up in arms at the lack of maintenance of the tower and its gradual deterioration with weakened mortar and loose bricks, falling tiles from the roof, and a similar neglect of the grounds which had become a dumping ground for people's garden rubbish. As a result of her article in the local newspaper, the town council was galvanized into action and undertook repairs (Julia Watson Rye News).

Aqueducts

The power of Roman technology and engineering during the age of the Roman Empire has left its dramatic mark across Europe, witnessed particularly in the striking infrastructure of aqueducts which stride across rural and urban landscapes from Britain to Spain and Greece. The first Roman Aqueduct, Aqua Appia was constructed in 312 BC to supply a water fountain at the city's cattle market; by the height of the Roman Empire in the third century AD the city had 11 aqueducts, sustaining a population of over a million. At this time their combined channel length is estimated to have been between 780 and 800 kilometres, of which 47 kilometres were carried above ground level, on masonry supports. Water was drawn from various springs and rivers around the city. Many of these Roman aqueducts have survived into the current period, and some are still in use. The construction of aqueducts has continued right up until the twentieth century—the Apulian aqueduct is an illustration of this (Photo 8.5).

In the earliest surviving architectural text *De Architectura*, Vitruvius writes of the materials selected for their construction but there is little said about how they were actually made. In the first century AD the Roman general Frontinus in *De Aquaeductu* wrote of the discrepancy

Photo 8.5 The Acquedotto in Puglia. Jessie Watson

between the amount of water taken into the aqueducts and the water supply, thus laying bear the stealing of water through illegal pipes inserted along the channels to divert water. Such exploitation of water power was endemic in the Roman Empire (Wikipedia). Roman aqueducts were a matter of civic pride and largely constructed to bring water into the towns and cities to meet the high levels of demand for the much-used public baths, fountains, ornamental gardens and private dwellings. The run off was used for the sewers. Gravity was the main principle for moving water through conduits made of stone, brick and concrete. Before the development of aqueduct technology, Romans, like most of their contemporaries in Europe had drunk water from natural sources such as springs and rivers, but many of the latter were polluted and carried disease.

Aqueducts in Italy are not only a feature of the Roman period. Many were built much later drawing on the same architectural and engineering principles that had been passed down from generation to generation. In the Southern Region of Puglia, water supply was a problem for centuries before the construction of the Apulian aqueduct at the end of the nineteenth century following National Reunification, when its necessity for economic and social development was recognized. Construction was started in 1906 and declared finished in 1939. This network now stretches 22,500 kilometres representing the largest aqueduct in Europe, incorporating 11,000 kilometres of sewage networks and 182 treatment plants.

The plan was ingenious since there were no viable rivers to channel in Puglia, engineers looked further north and tapped the headwaters of the Sele River in the mountains near Avellino in the Campania region, on the other (western) side of the Apennine watershed. The Sele flows down to the west for 64 kilometres to the Gulf of Salerno, but the aqueduct rerouted some of the water back across the watercourse and distributed it to the east through 1600 kilometres in the different channels of the aqueduct. A report in the New York Times in 1914 on the ongoing project wrote: '*[I]t is on a scale which gives it rank in the history of civilization as an ambitious project.*' The initial plan involved the construction of a long tunnel (15 km) through the Apennine range with a tunnel of 15 kilometres (9.4 miles) to bring water to the eastern part of the region by 1916 with at least 20,000 people working on the site. Though this was not realized, fresh water was brought to Bari by 1915.

The lay out of the Apulian aqueduct is complex, more like a web than a single channel, stretching 2189 kilometres (1360 miles) and passing through 99 channels and over 91 bridges, and providing water to more than four million inhabitants in the 258 cities, towns and villages across the Puglian region. Problems with maintenance, including those arising from the 1980 earthquake in the south of Italy, have meant that the system does not always function at ideal capacity. The aqueduct is managed by the *Acquedotto Pugliese* corporation that is involved in other hydrological projects in Puglia such as water purification plants, artificial catchment basins, artesian wells and desalinization plants.

Like many of the water infrastructures depicted here, its use has shifted and changed, with one section of the network, re-packaged as a walking and cycling route, reconstituting the channel as a site of aesthetics and leisure rather than the site of water supply. Once again we see how layers of recent history settle and re-settle upon one another

Brewery

During many decades, beer was the preferred alcohol of daily life in the US, whose production consumed large quantities of water. For the puritanical Protestants of the late nineteenth and early twentieth centuries widespread alcoholism and domestic violence became a matter of considerable concern. Coordinating their opposition with what they saw as the ill effects of alcohol, in alliance with the Women's Temperance Union and other Christian groups, they campaigned for the prohibition of the sale and consumption of alcohol. Between 1920 and 1933 as a result there was a national constitutional ban on the production and importation of alcohol—known as the Prohibition—which led to the widespread illegal production and consumption of alcohol across the states.

The effect of this ban on the material infrastructure of American cities was substantial—namely in the closure and subsequent decay of many of the breweries, a large proportion of which had been built by the various German communities that had emigrated to America. Many of these buildings lay empty until, like many other urban traces of a liquid past, they have been revived either as microbreweries with the new fashion for craft beers or redeveloped as private residences or offices for young urban professionals. Some of these historic buildings have gained heritage status being added to the National Register of Historic Places, like the Mission Brewery Plaza in the Middletown neighbourhood of San Diego (Photo 8.6).

Like many such buildings it had a multi-layered past serving first as an isolation hospital during the 1918 flu pandemic, then as an agar production plant from the 1920s to the 1980s when it was bought by a property development company and converted into offices, which then were lost to foreclosure in 1992, only to be later developed as offices and residential apartments in the following decade. Other developers have similarly restored old breweries—such as the long vacant Phoenix Brewery in Buffalo—often having to seek approval from the state historic preservation officials in the process. Part of the attraction of these buildings to the new cosmopolites are the original features where the exposed brick and duct work form an integral part of the conversion. As the planning board Vice Chair, Cynthia Schwartz put it: 'It is such a pleasure to see this building come back to life. It's just got such great bones. It's just been

Photo 8.6 Mission Brewery San Diego

waiting for someone to do the right thing with it,' she said. Thus, once again, we see urban materialities reformed and reshaped in a new context.

http://news.wbfo.org/post/plan-turn-old-brewery-loft-apartments-moves-forward-1

Bondi Outflow

On the north side of Bondi beach in Sydney in the midst of a municipal golf course a ventilation pipe stands above the cliffs like an obelisk adorning a town square. But this is no ornamental structure, it is the ventilation pipe for the Bondi Ocean Outfall Sewer (BOOS) which bears witness to early sewage infrastructures and which at various moments has assembled multiple publics in opposition to its presence. This pipe is connected to associated structures which include the sewers, vent shafts and pumping stations. An early work of colonial engineering (date), the first of its kind in Australia, is embedded in the socio-technical networks of the late nineteenth century, and now designated a significant heritage site. The BOOS is said to be a wonder of surveying accuracy for its time The Main Northern Ocean Outfall, or Bondi Ocean Outfall Sewer (BOOS), was the first ocean outfall sewer of its type to be designed and built in the country. It is one of the most significant engineering structures in Australia. It was a marvel of surveying accuracy for its time in that it allowed for the lining of the sewer before the tunnelling was completed. There are at least three stories to tell (Photo 8.7).

As a success of early engineering and construction methods the BOOS reduced the volume of polluted waters entering the Harbour and improved the health of the city's residents by moving polluted waters off shore, removing sewerage from the sewers, houses and water courses within the city, and drawing on improved ways of treating raw effluent. At a time when the city was growing fast it represented a major advance in the protection of public health of Sydney by reducing the discharge of sewage from inner city areas into Port Jackson. Its significance as an object representing the history of municipal facilities has led to its listing

Photo 8.7 The Bondi sewage outflow plant. Sophie Watson

by the State government on the State Government Inventory which designates heritage sites.

Like many traces of water objects in the city, this site is not a neutral artefact adorning the landscape, rather, it is a site of contestation, and also erasure of indigenous cultures, which has mobilized multiple publics in opposition. The first population to be affected by the sewerage process in Sydney had little power to resist. Before white settlement on Australia the Tank Stream, as it became known, was of cultural significance to the traditional owners of the Sydney cove area—the Gadigal people. Within 40 years of their arrival, the settlers had polluted the stream so badly that another water supply was sought. The stream was then used for sewage until the pollution reached such a high level that it had to be covered in stone providing the foundation for the city's growth. It was at this point that construction of the BOOS commenced to remove sewerage from the city, having in the meantime ruined the waters that were so significant to the original owners culture.

BOOS entered the public arena with force during the 1970s and 1980s. At this time much of the sewerage that entered the ocean was coming back onto the beaches, particularly Bondi beach making at that time a brown swirl that could be seen from planes flying in and out of Sydney and carrying poo into the waves of the surfers on the beach. In the late 1980s a group of activists set up an organization to raise awareness of the issue and to assemble opposition to the sewerage treatment process. A new matter of concern had arisen with a public voice in the media and political sphere. Assembling a multiplicity of resisters—locals, environmentalists, trade unions and others—in 1989 they organized the Turn Back the Tide concert on Bondi Beach where a quarter of a million people articulated their concern. As Hawkins points out (2004) this was not a pre-existing public, rather it was one formed around an issue brought together through a shared anxiety and disgust and deploying innovative discourses of resistance such as 'cartoons depicting people diving into toilet bowls or lifesavers hauling a roll of toilet paper into the surf instead of a rescue rope' to make their point. Something deeply disturbing was at stake here in the breaking down of the boundaries between public and private, as bodily matters—excrement, which are typically kept in place and out of sight, entered the public sphere in a dramatic manner (Hawkins 2011).

References

Benjamin, Walter. 1982. *Arcades Project*. London: Bloomsbury.
Edensor, Tim. 2005. *Industrial Ruins: Aesthetics, Materiality and Memory*. Oxford: Berg.
Hawkins, Gay. 2004. Shit in Public. *Australian Humanities Review* 31 (2): 33–47.
———. 2011. Commentary. *Environment and Planning A* 43: 2001–2006.
Hetherington, K. 2014. Museums and the 'Death of Experience': Singularity, Interiority and the Outside. *International Journal of Heritage Studies* 20 (1): 72–85.
de Laet, Marianne, and Anne Marie Mol. 2000. The Zimbabwe Bush Pump: Mechanics of a Fluid Technology. *Social Studies of Science* 30 (2): 225–263.

9

Final Word

This is a book that has aimed to bring to light the significance of water to everyday life in cities and a multiplicity of practices which often go unnoticed or barely remarked. Water in cities is as much about the passions of those that swim in its hidden places or sit by its fountains, as it is about the pipes that deliver it, or institutional interventions. Water is as much a cultural object as a material substance—a source of complex meanings and practices which are constituted by, and constitutive of, the different political and economic structures of water provision in particular places. No two cities are the same, and their historical, socio-cultural and political/economic specificities have implications for the water that flows through them. Water exists as a resource through a complex intersection of socio-technical networks and systems underlining the very entanglements of humans with the environment and nature.

As fluid and vibrant matter water defies containment and definition, flowing literally and metaphorically into every part of life. This book has aimed to break out of traditional accounts of water, following water as it emerges, shifts and settles in cities in different ways. I have explored different elements of water's presence, as material object, cultural representation, as movement, as actor, as practice and as ritual, across the chapters, to highlight its importance beyond the quenching of thirst, while recognizing also that in some parts of the world, even this possibility is the privilege of the few. I have also drawn attention to the power of water in assembling and differentiating heterogeneous publics, for work, for pleasure, for governance and for consumption. In this we see that the publics and differences assembled are themselves not fixed, but shift and change across time and place. We have seen also how water is imbricated in relations of power in all its rhythms and forms.

The different chapters of the book have explored how water sets in train a plethora of technologies for producing and directing it amidst a complexity of governmental and regulatory practices as well as private markets and forms of provision. In cities of the Global North, water only becomes a matter of concern in times of scarcity or drought, or through its potential for danger, or when the accumulation of profit by water providers is threatened. Yet this luxury of ignoring the impending crisis regarding the availability of water, its pollution or its overabundance in floods will no longer be the prerogative of those living in the wealthier parts of the world. These problems will extend beyond the cities of the Global South to cities from the US to New Zealand and be impossible to ignore. Over the course of the twenty-first century, as the effects of climate change will be more keenly felt, many more people will experience these impacts also. Migration, once fuelled by economic or political uncertainties, or war, will be one of the major consequences. Fighting for scarce resources within and between countries may well be another. This book has shown how water seeps into every aspect of everyday life in the city. Its future should concern us all.